Lecture Notes in Computer Science 9541

Commenced Publication in 1973
Founding and Former Series Editors:
Gerhard Goos, Juris Hartmanis, and Jan van Leeuwen

More information about this series at http://www.springer.com/series/7407

Mohammad Taghi Hajiaghayi
Mohammad Reza Mousavi (Eds.)

Topics in Theoretical Computer Science

The First IFIP WG 1.8 International Conference, TTCS 2015
Tehran, Iran, August 26–28, 2015
Revised Selected Papers

 Springer

Editors
Mohammad Taghi Hajiaghayi
Computer Science Department
University of Maryland
College Park, MD
USA

Mohammad Reza Mousavi
Centre for Research on Embedded Systems
Halmstad University
Halmstad
Sweden

ISSN 0302-9743   ISSN 1611-3349   (electronic)
Lecture Notes in Computer Science
ISBN 978-3-319-28677-8   ISBN 978-3-319-28678-5   (eBook)
DOI 10.1007/978-3-319-28678-5

Library of Congress Control Number: 2015958862

LNCS Sublibrary: SL1 – Theoretical Computer Science and General Issues

Printed on acid-free paper

This Springer imprint is published by SpringerNature
The registered company is Springer International Publishing AG Switzerland

Preface

Welcome to the proceedings of the First IFIP International Conference on Topics in Theoretical Computer Science (TTCS 2015)!

This volume contains the revised papers presented at TTCS 2015. The conference was held during August 26–28, 2015, at the Institute for Research in Fundamental Sciences (IPM), Tehran, Iran.

For this first edition of TTCS, we received 48 submissions from 12 different countries. An international Program Committee comprising 45 leading scientists from 14 countries reviewed the papers thoroughly providing on average five review reports for each paper. We ended up accepting 11 submissions, which translates into less than 23 % of all submissions. This means that the process was highly selective and only very high quality papers were accepted.

The program also included three invited talks by the following world-renowned computer scientists:

- Prof. Anuj Dawar, Cambridge University, UK,
- Prof. Michael Fellows, Charles Darwin University, Australia, and
- Dr. Mehrnoosh Sadrzadeh, Queen Mary University of London, UK.

These invited lectures are also represented either by an abstract or by a full paper in the proceedings.

Additionally, the program featured four talks in the PhD Forum, which are not included in the proceedings.

We thank IPM, and in particular the Organizing Committee, for having provided various facilities and for their generous support. We believe that, thanks to their help, the conference organization was run smoothly and conveniently. We are also grateful to our program committee for their professional and hard work in providing expert review reports and thorough discussions leading to a very interesting and strong program. We also acknowledge the excellent facilities provided by the EasyChair system, which have been crucial in managing the process of submission, selection, revision, and publication of the manuscripts included in the proceedings.

November 2015

Mohammad Taghi Hajiaghayi
Mohammad Reza Mousavi

Organization

Program Committee

Mohammad A. Abam	Sharif University of Technology, Tehran, Iran
Saeed Akbari	Sharif University of Technology, Tehran, Iran
Saeed Alaei	Cornell University, USA
Farhad Arbab	CWI and Leiden University, The Netherlands
S. Arun-Kumar	Indian Institue of Technology Delhi, India
Mohammad H. Bateni	Google Research, USA
Salman Beigi	IPM, Iran
Ilaria Castellani	Inria Sophia Antipolis Mediterranee, France
Dave Clarke	Uppsala University, Sweden
Pieter Cuijpers	Eindhoven University of Technology, The Netherlands
Amir Daneshgar	Sharif University of Technology, Iran
Fedor Fomin	University of Bergen, Norway
Fatemeh Ghassemi	University of Tehran, Iran
Ali Ghodsi	University of Waterloo, Canada
Mohammad Ghodsi	Sharif University of Technology, Iran
Mohammad T. Hajiaghayi	University of Maryland, USA
Ichiro Hasuo	University of Tokyo, Japan
Matthew Hennessy	Trinity College Dublin, Ireland
Nicole Immorlica	Microsoft Research, USA
Mohammad M. Jaghoori	AMC of University of Amsterdam, The Netherlands
Amin Karbasi	Yale University, USA
Jeroen Keiren	Open University Netherlands, The Netherlands
Amit Kumar	Indian Institute of Technology Delhi, India
Bas Luttik	Eindhoven University of Technology, The Netherlands
Mohammad Mahdian	Google Research, USA
Hamid Mahini	University of Maryland, USA
Mohammad Mahmoody	University of Virginia, USA
Jose Meseguer	University of Illinois, USA
Vahab Mirrokni	Google Research, USA
Bojan Mohar	Simon Fraser University, Canada
Morteza Monemizadeh	Frankfurt University, Germany
Lary Moss	Indiana University, USA
Mohammad R. Mousavi	Halmstad University, Sweden
Shayan Oveisgharan	University of Washington, USA
Jun Pang	University of Luxembourg, Luxembourg
Debmalya Panigrahi	Duke University, USA
Jorg Sack	Carleton University, Canada

Gerardo Schneider	Chalmers — University of Gothenburg, Sweden
Mohit Singh	Microsoft Research, USA
Marjan Sirjani	Reykjavik University, Reykjavik, Iceland
Dimitrios M. Thilikos	CNRS, France and University of Athens, Greece
S. Venkatasubramanian	University of Utah, USA
Walter Vogler	Augsburg University, Germany
Jan Vondrak	IBM Almaden Research Center, USA
Tim Willemse	Eindhoven University of Technology, The Netherlands

Additional Reviewers

Abbasi Zadeh, Sepehr
Alégroth, Emil
Baharifard, Fatemeh
Banijamali, Ershad
Beohar, Harsh
Bujtor, Ferenc
Chen, Lin
Clairambault, Pierre
Corby, Olivier
Damaschke, Peter
Deng, Yuxin

Fox, Kyle
Gacs, Peter
Haney, Samuel
Homapour, Hamid
Jonker, Hugo
Kell, Nathaniel
Khamespanah, Ehsan
Krčál, Jan
Langetepe, Elmar
Melliès, Paul-André
Nemati, Soheil

Sabahi Kaviani, Zeynab
Schmaltz, Julien
Shameli, Seyed Ali
Shariatpanahi,
   Seyed Pooya
Sidorova, Natalia
Suykens, Johan
Tichy, Matthias
Visser, Arnoud
Volpato, Michele

Abstracts of Invited Talks

New Directions in Parameterized Algorithmics

Michael Fellows

Charles Darwin University, Australia

The talk will review some basics of the field, and then focus on some new directions in parameterized/multivariate algorithmics. These include:

- The systematic deconstruction of NP-hardness results for problems unrealistically legislated with real numbers, by parameterizing on the size of a relevant finitized arithmetic system.
- Fresh paradigms for deploying parameterization to FPT-turbocharge heuristics (such as greedy algorithms) and other subroutines of current approaches in practical computing for NP-hard problems.
- Aggressive aggregate parameterization including generative parameterization of typical instances of hard problems, building on and deepening the parameter ecology program.
- Parameterization in the context of *dynamic problems* where inputs change (a bit) and solutions need to be changed (a bit).
- The axiomatization of *groovy FPT*, where canonically structured kernelization is canonically convertible to: P-time approximation algorithms; inductive gradients for local search; sharper turbocharging for greedy algorithms, local search and genetic recombination heuristics — opening up a whole new level of *groovy* lower bound questions, such as the existence of *groovy polynomial kernels*. (Recent results show that some FPT parameterized problems admit polynomial kernels, but do not admit groovy polynomial kernelization unless P = NP.)
- The accidental origins of parameterized complexity in graph minor theory (well quasi-ordering + FPT order tests) is a general phenomenon: it has recently been shown that a parameterized problem is FPT if and only if this can be derived from a well-behaved WQO context. Where this leads is entirely open.

Distributional Sentence Entailment Using Density Matrices

Esma Balkr[1], Mehrnoosh Sadrzadeh[1], and Bob Coecke[2]

[1] Queen Mary University of London
[2] University of Oxford

Abstract. Categorical compositional distributional model of Clark, Coecke, and Sadrzadeh suggests a way to combine grammatical composition of the formal, type logical models with the corpus based, empirical word representations of distributional semantics. This paper contributes to the project by expanding the model to also capture entailment relations. This is achieved by extending the representations of words from points in meaning space to density operators, which are probability distributions on the subspaces of the space. A symmetric measure of similarity and an asymmetric measure of entailment is defined, where lexical entailment is measured using von Neumann entropy, the quantum variant of Kullback-Leibler divergence. Lexical entailment, combined with the composition map on word representations, provides a method to obtain entailment relations on the level of sentences. Truth theoretic and corpus-based examples are provided.

On Symmetric and Choiceless Computation

Anuj Dawar

University of Cambridge Computer Laboratory, William Gates Building,
J.J. Thomson Avenue, Cambridge, CB3 0FD, UK
anuj.dawar@cl.cam.ac.uk

Formal models of computation such as Turing machines are usually defined as performing operations on strings of symbols. Indeed, for most purposes, it suffices to consider strings over a two-letter alphabet $\{0, 1\}$. Decision problems are defined as sets of strings, and complexity classes as sets of decision problems. However, many natural algorithms are described on more abstract structure (such as graphs) because this is the natural level of abstraction at which to describe the problem being solved. Of course, we know that the abstract structures can be ultimately represented as strings (and, indeed, have to be in actual computational devices), but the representation comes at a cost. The same abstract structure may have many different string representations and the implementation of the algorithm may break the intended abstraction.

Research in the area of finite model theory and descriptive complexity (see [11, 13]) has, over the years, developed a number of techniques of describing algorithms and complexity classes directly on classes of relational structures, rather than strings. Along with this, many methods of proving inexpressiblity results have been shown, often described in terms of games. A key question that has been the focus of this research effort is whether the complexity class P admits a descriptive characterisation (see [10, Chap. 11]).

A recent paper [1] ties some of the logics studied in finite model theory to natural circuit complexity classes, and shows thereby that inexpressibility results obtained in finite model theory can be understood as lower bound results on such classes. In this presentation, I develop the methods for proving lower bound results in the form of combinatorial arguments on circuits, without reference to logical definability. The present abstract gives a brief account of the results and methods.

Contents

Distributional Sentence Entailment
Using Density Matrices

Esma Balkir[1], Mehrnoosh Sadrzadeh[1](✉), and Bob Coecke[2]

[1] Queen Mary University of London, London, UK
m.sadrzadeh@qmul.ac.uk
[2] University of Oxford, Oxford, UK

Abstract. Categorical compositional distributional model of Clark, Coecke, and Sadrzadeh suggests a way to combine grammatical composition of the formal, type logical models with the corpus based, empirical word representations of distributional semantics. This paper contributes to the project by expanding the model to also capture entailment relations. This is achieved by extending the representations of words from points in meaning space to density operators, which are probability distributions on the subspaces of the space. A symmetric measure of similarity and an asymmetric measure of entailment is defined, where lexical entailment is measured using von Neumann entropy, the quantum variant of Kullback-Leibler divergence. Lexical entailment, combined with the composition map on word representations, provides a method to obtain entailment relations on the level of sentences. Truth theoretic and corpus-based examples are provided.

1 Introduction

The term *distributional semantics* is almost synonymous with the term *vector space models of meaning*. This is because vector spaces are natural candidates for modelling the *distributional hypothesis* and contextual similarity between words [11]. In a nutshell, this hypothesis says that words that often occur in the same contexts have similar meanings. So for instance, 'ale' and 'lager' are similar since they both often occur in the context of 'beer', 'pub', and 'pint'. The obvious practicality of these models, however, does not guarantee that they possess the expressive power needed to model all aspects of meaning. Current distributional models mostly fall short of successfully modelling subsumbtion and entailment [19]. There are a number of models that use distributional similarity to enhance textual entailment [4,13]. However, most of the work from the distributional semantics community has been focused on developing more sophisticated metrics on vector representations [17,20,27].

In this paper we suggest the use of density matrices instead of vector spaces as the basic distributional representations for the meanings of words. Density matrices are widely used in quantum mechanics, and are a generalization of vectors. There are several advantages to using density matrices to model meaning.

© IFIP International Federation for Information Processing 2016
Published by Springer International Publishing Switzerland 2016. All Rights Reserved.
M.T. Hajiaghayi and M.R. Mousavi (Eds.): TTCS 2015, LNCS 9541, pp. 1–22, 2016.
DOI: 10.1007/978-3-319-28678-5_1

Firstly, density matrices have the expressive power to represent all the information vectors can represent: they are a suitable implementation of the distributional hypothesis. They come equipped with a measure of information content, and so provide a natural way of implementing asymmetric relations between words such as hyponymy-hypernymy relations. Futhermore, they form a compact closed category. This allows the previous work of [7,10] on obtaining representations for meanings of sentences from the meaning of words to be applicable to density matrices. The categorical map from meanings of words to the meaning of the sentence respects the order induced by the relative entropy of density matrices. This promises, given suitable representations of individual words, a method to obtain entailment relations on the level of sentences, inline with the lexical entailment of *natural logic*, e.g., see [21], rather than the traditional logical entailment of Montague semantics.

Related Work. This work builds upon and relates to the literature on compositional distributional models, distributional lexical entailment, and the use of density matrices in computational linguistics and information retrieval.

There has been a recent interest in methods of composition within the distributional semantics framework. There are a number of composition methods in literature. See [14] for a survey of compositional distributional models and a discussion of their strengths and weaknesses. This work extends the work presented in [7,10], a compositional model based on category theory. Their model was shown to outperform the competing compositional models in [15].

Research on distributional entailment has mostly been focused on lexical entailment. One notable exception is the work in [3], which uses the distributional data on adjective-noun and quantifier-noun pairs to train a classifier; the results are then utilized to detect novel noun pairs that have the same relation. There are a number of non-symmetric lexical entailment measures, e.g., see [8,17,19,27], all of which rely on some variation of the *Distributional Inclusion Hypothesis*: "If u is semantically narrower than v, then a significant number of salient distributional features of u are also included in the feature vector of v" [17]. In their experiments, the authors of [12] show that while if a word v entails another word w then the characteristic features of v is a subset of the ones for w, it is not necessarily the case that the inclusion of the characteristic features v in w indicate that v entails w. One of their suggestions for increasing the prediction power of their method is to include more than one word in the features.

The work presented in [23] uses a measure based on entropy to detect hyponym-hypernym relationships in given pairs. The measure they suggest rely on the hypothesis that hypernyms are semantically more general than hyponyms, and therefore tend to occur in less informative contexts. The authors of [16] rely on a very similar idea, and use KL-divergence between the target word and the basis words to quantify the semantic content of the target word. They conclude that this method performs equally well in detecting hyponym-hypernym pairs as their baseline prediction method that only considers the overall frequency of the word in corpus. They reject the hypothesis that more general words occur in less informative contexts. Their method differs from ours in that they use relative entropy to quantify the overall information content of a word, and not to compare two target words to each other.

The work presented in [22] extend the compositional model of [7,10] to include density matrices as we do, but use it for modeling homonymy and polysemy. Their approach is complementary to ours, and in fact, they show that it is possible to merge the two constructions. The work presented in [6] uses density matrices to model context effects in a conceptual space. In their quantum mechanics inspired model, words are represented by mixed states and each eigenstate represents a sense of the word. Context effects are then modelled as quantum collapse. The authors in [5] use density matrices to encode dependency neighbourhoods, with the aim of modelling context effects in similarity tasks; the work presented in [26] uses density matrices to sketch out a theory of information retrieval, and connects the logic of the space and of density matrices via an order relation that makes the set of projectors in a Hilbert space into a complete lattice. They then use this order to define an entailment relation. Finally, the work presented in [25] shows that using density matrices to represent documents provides significant improvement on realistic IR tasks.

This paper is based on the MSc Thesis of the first author [2].

2   Background

Definition 1. *A monoidal category is **compact closed** if for any object A, there are left and right dual objects, i.e. objects A^r and A^l, and morphisms $\eta^l : I \to A \otimes A^l$, $\eta^r : I \to A^r \otimes A$, $\epsilon^l : A^l \otimes A \to I$ and $\epsilon^r : A \otimes A^r \to I$ that satisfy:*

$$(1_A \otimes \epsilon^l) \circ (\eta^l \otimes 1_A) = 1_A \quad (\epsilon^r \otimes 1_A) \circ (1_A \otimes \eta^r) = 1_A$$
$$(\epsilon^l \otimes 1_{A^l}) \circ (1_{A^l} \otimes \eta^l) = 1_{A^l} \quad (1_{A^r} \otimes \epsilon^r) \circ (\eta^r \otimes 1_{A^r}) = 1_{A^r}$$

Compact closed categories are used to represent *correlations*, and in categorical quantum mechanics they model maximally entangled states. [1] The η and ϵ maps are useful in modeling the interactions of the different parts of a system. To see how this relates to natural language, consider a simple sentence with an object, a subject and a transitive verb. The meaning of the entire sentence is not simply an accumulation of the meanings of its individual words, but depends on how the transitive verb relates the subject and the object. The η and ϵ maps provide the mathematical formalism to specify such interactions. The distinct left and right duals ensure that compact closed categories can take word order into account.

There is a graphical calculus used to reason about monoidal categories [9]. In the graphical language, objects are wires, and morphisms are boxes with incoming and outgoing wires of types corresponding to the input and output types of the morphism. The identity object is depicted as empty space, so a state $\psi : I \to A$ is depicted as a box with no input wire and an output wire with type A. The duals of states are called *effects*, and they are of type $A \to I$. Let

$f : A \to B$, $g : B \to C$ and $h : C \to D$, and $1_A : A \to A$ the identity function on A. 1_A, f, $f \otimes h$, $g \circ f$ are depicted as follows:

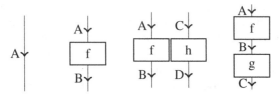

The state $\psi : I \to A$, the effect $\pi : A \to I$, and the scalar $\psi \circ \pi$ are depicted as follows:

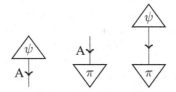

The maps $\eta^l, \eta^r, \epsilon^l$ and ϵ^r take the following forms in the graphical calculus:

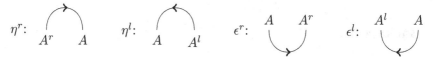

The axioms of compact closure, referred to as the *snake identities* because of the visual form they take in the graphical calculus, are represented as follows:

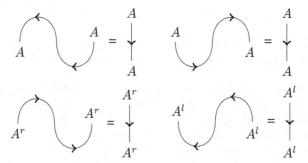

More generally, the reduction rules for diagrammatic calculus allow continuous deformations. One such deformation we will make use of is the *swing rule*:

Definition 2 [18]. *A **pregroup** $(P, \leq, \cdot, 1, (-)^l, (-)^r)$ is a partially ordered monoid in which each element a has both a left adjoint a^l and a right adjoint a^r such that $a^l a \leq 1 \leq aa^l$ and $aa^r \leq 1 \leq a^r a$.*

If $a \leq b$ it is common practice to write $a \rightarrow b$ and say that a reduces to b. This terminology is useful when pregroups are applied to natural language, where each word gets assigned a pregroup type freely generated from a set of basic elements. The sentence is deemed to be grammatical if the concatenation of the types of the words reduce to the simple type of a sentence. For example reduction for a simple transitive sentence is $n(n^r s n^l)n \rightarrow 1\, sn^l n \rightarrow 1\, s1 \rightarrow s$.

A pregroup $\mathbf{P}$ is a concrete instance of a compact closed category. The $\eta^l, \eta^r, \epsilon^l, \epsilon^r$ maps are $\eta^l = [1 \leq p \cdot p^l]$, $\epsilon^l = [p^l \cdot p \leq 1]$, $\eta^r = [1 \leq p^r \cdot p]$, $\epsilon^r = [p \cdot p^r \leq 1]$.

FVect as a Concrete Compact Closed Category. Finite dimensional vector spaces over the base field $\mathbb{R}$, together with linear maps form a monoidal category, referred to as **FVect**. The monoidal tensor is the usual vector space tensor and the monoidal unit is the base field $\mathbb{R}$. It is also a compact closed category where $V^l = V^r = V$. The compact closed maps are defined as follows:

Given a vector space V with basis $\{\overrightarrow{e_i}\}_i$,

$$\eta_V^l = \eta_V^r : \mathbb{R} \rightarrow V \otimes V \qquad\qquad \epsilon_V^l = \epsilon_V^r : V \otimes V \rightarrow \mathbb{R}$$

$$1 \mapsto \sum_i \overrightarrow{e_i} \otimes \overrightarrow{e_i} \qquad\qquad \sum_{ij} c_{ij}\, \overrightarrow{v_i} \otimes \overrightarrow{w_i} \mapsto \sum_{ij} c_{ij} \langle \overrightarrow{v_i} | \overrightarrow{w_i} \rangle$$

Categorical Representation of Meaning Space. The tensor in **FVect** is commutative up to isomorphism. This causes the left and the right adjoints to be the same, and thus the left and the right compact closed maps to coincide. Thus **FVect** by itself cannot take the effect of word ordering on meaning into account. [7,10] propose a way around this obstacle by considering the product category **FVect** $\times$ **P** where **P** is a pregroup.

Objects in **FVect** $\times$ **P** are of the form (V, p), where V is the vector space for the representation of meaning and p is the pregroup type. There exists a morphism $(f, \leq) : (V, p) \rightarrow (W, q)$ if there exists a morphism $f : V \rightarrow W$ in **FVect** and $p \leq q$ in **P**.

The compact closed structure of **FVect** and **P** lifts componentwise to the product category **FVect** $\times$ **P**:

$$\eta^l : (\mathbb{R}, 1) \rightarrow (V \otimes V, p \cdot p^l) \qquad\qquad \eta^r : (\mathbb{R}, 1) \rightarrow (V \otimes V, p^r \cdot p)$$

$$\epsilon^l : (V \otimes V, p^l \cdot p) \rightarrow (\mathbb{R}, 1) \qquad\qquad \epsilon^r : (V \otimes V, p \cdot p^r) \rightarrow (\mathbb{R}, 1).$$

Definition 3. *An object (V, p) in the product category is called a **meaning space**, where V is the vector space in which the meanings $\overrightarrow{v} \in V$ of strings of type p live.*

Definition 4. *From-Meanings-of-Words-to-the-Meaning-of-the-Sentence Map. Let $v_1 v_2 \ldots v_n$ be a string of words, each v_i with a meaning space representation $\overrightarrow{v_i} \in (V_i, p_i)$. Let $x \in P$ be a pregroup type such that $[p_1 p_2 \ldots p_n \leq x]$. Then the meaning vector for the string is $\overrightarrow{v_1 v_2 \ldots v_n} := f(\overrightarrow{v_1} \otimes \overrightarrow{v_2} \otimes \ldots \otimes \overrightarrow{v_n}) \in (W, x)$, where f is defined to be the application of the compact closed maps* *obtained from the reduction* $[p_1 p_2 \ldots p_n \leq x]$ *to the composite vector space* $V_1 \otimes V_2 \otimes \ldots \otimes V_n$.

This framework uses the maps of the pregroup reductions and the elements of objects in **FVect**. The diagrammatic calculus provides a tool to reason about both. As an example, take the sentence "John likes Mary". It has the pregroup type $nn^r sn^l n$, and the vector representations $\overrightarrow{John}, \overrightarrow{Mary} \in V$ and $\overrightarrow{likes} \in V \otimes S \otimes V$. The morphism in **FVect** $\times$ **P** corresponding to the map defined in Definition 4 is of type $(V \otimes (V \otimes S \otimes V) \otimes V, nn^r sn^l n) \to (S, s)$. From the pregroup reduction $[nn^r sn^l n \to s]$ we obtain the compact closed maps $\epsilon^r 1 \epsilon^l$. In **FVect** this translates into $\epsilon_V \otimes 1_S \otimes \epsilon_V : V \otimes (V \otimes S \otimes V) \otimes V \to S$. This map, when applied to $\overrightarrow{John} \otimes \overrightarrow{likes} \otimes \overrightarrow{Mary}$, has the following depiction in the diagrammatic calculus:

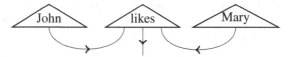

Note that this construction treats the verb 'likes' essentially as a relation that takes two inputs of type V, and outputs a vector of type S. For the explicit calculation, note that $\overrightarrow{likes} = \sum_{ijk} c_{ijk} \overrightarrow{v_i} \otimes \overrightarrow{s_j} \otimes \overrightarrow{v_k}$, where $\{\overrightarrow{v_i}\}_i$ is an orthonormal basis for V and $\{\overrightarrow{s_j}\}_j$ is an orthonormal basis for S. Then

$$\overrightarrow{John \; likes \; Mary} = \epsilon_V \otimes 1_S \otimes \epsilon_V (\overrightarrow{John} \otimes \overrightarrow{likes} \otimes \overrightarrow{Mary}) \tag{1}$$

$$= \sum_{ijk} \langle \overrightarrow{John} | \overrightarrow{v_i} \rangle \overrightarrow{s_j} \langle \overrightarrow{v_k} | \overrightarrow{Mary} \rangle \tag{2}$$

The reductions in diagrammatic calculus help reduce the final calculation to a simpler term. The non-reduced reduction, when expressed in dirac notation is $(\langle \epsilon_V^r | \otimes 1_S \otimes \langle \epsilon_V^l |) \circ | \overrightarrow{John} \otimes \overrightarrow{likes} \otimes \overrightarrow{Mary} \rangle$. But we can *swing* $\overrightarrow{John}$ and $\overrightarrow{Mary}$ in accord with the reduction rules in the diagrammatic calculus. The diagram then reduces to:

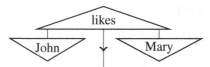

This results in a simpler expression that needs to be calculated: $(\langle \overrightarrow{John} | \otimes 1_S \otimes \langle \overrightarrow{Mary} |) \circ | \overrightarrow{likes} \rangle$.

3 Density Matrices as Elements of a Compact Closed Category

Recall that in **FVect**, vectors $| \overrightarrow{v} \rangle \in V$ are in one-to-one correspondence with morphisms of type $v : I \to V$. Likewise, pure states of the form $| \overrightarrow{v} \rangle \langle \overrightarrow{v} |$ are in one-to-one correspondence with morphisms $v \circ v^\dagger : V \to V$ such that $v^\dagger \circ v = \mathrm{id}_I$, where $v^\dagger$ denotes the adjoint of v (notice that this corresponds to the condition that $\langle v|v \rangle = 1$). A general (mixed) state ρ is a positive morphism of the form $\rho : V \to V$. One can re-express the mixed states $\rho : V \to V$ as elements $\rho : I \to V^* \otimes V$. Here $V^* = V^l = V^r = V$.

Definition 5. f *is a* ***completely positive map*** *if* f *is positive for any positive operator* A, *and* ($\mathrm{id}_V \otimes f)B$ *is positive for any positive operator* B *and any space* V.

Completely positive maps in **FVect** form a monoidal category [24]. Thus one can define a new category **CPM(FVect)** where the objects of **CPM(FVect)** are the same as those of **FVect**, and morphisms $A \to B$ in **CPM(FVect)** are completely positive maps $A^* \otimes A \to B^* \otimes B$ in **FVect**. The elements $I \to A$ in **CPM(FVect)** are of the form $I^* \otimes I \to A^* \otimes A$ in **FVect**, providing a monoidal category with density matrices as its elements.

CPM(FVect) in Graphical Calculus. A morphism $\rho : A \to A$ is positive if and only if there exists a map $\sqrt{\rho}$ such that $\rho = \sqrt{\rho}^\dagger \circ \sqrt{\rho}$. In **FVect**, the isomorphism between $\rho : A \to A$ and $\ulcorner \rho \urcorner : I \to A^* \otimes A$ is provided by $\eta^l = \eta^r$. The graphical representation of ρ in **FVect** then becomes:

Notice that the categorical definition of a positive morphism coincides with the definition of a positive operator in a vector space, where $\sqrt{\rho}$ is the square root of the operator.

The graphical depiction of completely positive morphisms come from the following theorem:

Theorem 1 (*Stinespring Dilation Theorem*). $f : A^* \otimes A \to B^* \otimes B$ *is completely positive if and only if there is an object* C *and a morphism* $\sqrt{f} : A \to C \otimes B$ *such that the following equation holds:*

$\sqrt{f}$ and C here are not unique. For the proof of the theorem see [24].

Theorem 2. CPM(FVect) *is a compact closed category where as in* **FVect**, *$V^r = V^l = V$ and the compact closed maps are defined to be:*

$$\eta^l = (\eta^r_V \otimes \eta^l_V) \circ (1_V \otimes \sigma \otimes 1_V) \qquad \eta^r = (\eta^l_V \otimes \eta^r_V) \circ (1_V \otimes \sigma \otimes 1_V)$$

$$\epsilon^l = (1_V \otimes \sigma \otimes 1_V) \circ (\epsilon^r_V \otimes \epsilon^l_V) \qquad \epsilon^r = (1_V \otimes \sigma \otimes 1_V) \circ (\epsilon^l_V \otimes \epsilon^r_V)$$

where σ is the swap map defined as $\sigma(v \otimes w) = (w \otimes v)$.

Proof. The graphical construction of the compact closed maps boils down to doubling the objects and the wires. The identities are proved by adding bends in the wires. Consider the diagram for η^r:

These maps satisfy the axioms of compact closure since the components do.

The concrete compact closed maps are as follows:

$$\eta^l = \eta^r : \mathbb{R} \to (V \otimes V) \otimes (V \otimes V)$$

$$::1 \mapsto \sum_i \overrightarrow{e_i} \otimes \overrightarrow{e_i} \otimes \sum_j \overrightarrow{e_j} \otimes \overrightarrow{e_j}$$

$$\epsilon^l = \epsilon^r : (V \otimes V) \otimes (V \otimes V) \to \mathbb{R}$$

$$:: \sum_{ijkl} c_{ijkl} \; \overrightarrow{v_i} \otimes \overrightarrow{w_j} \otimes \overrightarrow{u_k} \otimes \overrightarrow{p_l} \mapsto \sum_{ijkl} c_{ijkl} \; \langle \overrightarrow{v_i} | \overrightarrow{u_k} \rangle \langle \overrightarrow{w_j} | \overrightarrow{p_l} \rangle$$

Let $\rho : V_1 \otimes V_2 \otimes \ldots \otimes V_n \to V_1 \otimes V_2 \otimes \ldots \otimes V_n$ be a density operator defined on an arbitrary composite space $V_1 \otimes V_2 \otimes \ldots \otimes V_n$. Then it has the density matrix representation $\rho : I \to (V_1 \otimes V_2 \otimes \ldots \otimes V_n)^* \otimes (V_1 \otimes V_2 \otimes \ldots \otimes V_n)$. Since the underlying category **FVect** is symmetric, it has the swap map σ. This provides us with the isomorphism:

$$(V_1 \otimes V_2 \otimes \ldots \otimes V_n)^* \otimes (V_1 \otimes V_2 \otimes \ldots \otimes V_n) \sim (V_1^* \otimes V_1) \otimes (V_2^* \otimes V_2) \otimes \ldots \otimes (V_n^* \otimes V_n)$$

So ρ can be equivalently expressed as $\rho : I \to (V_1^* \otimes V_1) \otimes (V_2^* \otimes V_2) \otimes \ldots \otimes (V_n^* \otimes V_n)$. With this addition, we can simplify the diagrams used to express density matrices by using a single thick wire for the doubled wires. Doubled compact closed maps can likewise be expressed by a single thick wire.

The diagrammatic expression of a from-meanings-of-words-to-the-meaning-of-the-sentence map using density matrices will therefore look exactly like the depiction of it in **FVect**, but with thick wires.

4   Using Density Matrices to Model Meaning

If one wants to use the full power of density matrices in modelling meaning, one needs to establish an interpretation for the distinction between *mixing* and *superposition* in the context of linguistics. Let *contextual features* be the salient, quantifiable features of the contexts a word is observed in. Let the basis of the space be indexed by such contextual features. Individual *contexts*, such as words in an *n*-word window of a text, can be represented as the superposition of the bases corresponding to the contextual features observed in it. So each context corresponds to a pure state. Words are then probability distributions over the contexts they appear in. The simple co-occurrence model can be cast as a special case of this more general approach, where features and contexts are the same. Then all word meanings are mixtures of basis vectors, and they all commute with each other.

Similarity for Density Matrices. Fidelity is a good measure of similarity between two density matrix representations of meaning because of its properties listed below.

Definition 6. *The fidelity of two density operators ρ and σ is $F(\rho, \sigma) := tr\sqrt{\rho^{1/2}\sigma\rho^{1/2}}$.*

Some useful properties of fidelity are:

1. $F(\rho, \sigma) = F(\sigma, \rho)$.
2. $0 \leq F(\rho, \sigma) \leq 1$.
3. $F(\rho, \sigma) = 1$ if and only if $\rho = \sigma$.
4. If $|\phi\rangle\langle\phi|$ and $|\psi\rangle\langle\psi|$ are two pure states, their fidelity is equal to $|\langle\phi|\psi\rangle|$.

These properties ensure that if the representations of two words are not equal to each other, they will not be judged perfectly similar, and, if two words are represented as projections onto one dimensional subspaces, their similarity value will be equal to the usual cosine similarity of the vectors.

Entailment for Density Matrices. To develop a theory of entailment using density matrices as the basic representations, we assume the following hypothesis:

Definition 7 (Distributional Hypothesis for Hyponymy). *The meaning of a word w subsumes the meaning of a word v if and only if it is appropriate to use w in all the contexts v is used.*

This is a slightly more general version of the *Distributional Inclusion Hypothesis* (DIH) stated in [17]. The difference lies in the additional power the density matrix formalism provides: the distinction between mixing and superposition. Further, DIH only considers whether or not the target word occurs together with the salient distributional feature at all, and ignores any possible statistically significant correlations of features; here again, the density matrix formalism offers a solution.

Note that [12] show that while there is ample evidence for the distributional inclusion hypothesis, this in itself does not necessarily provide a method to detect hyponymy-hypernymy pairs. One of their suggestions for improvement is to consider more than one word in the features, equivalent to what we do here by taking correlations into account in a co-occurrence space where the bases are context words.

Relative entropy quantifies the distinguishability of one distribution from another. The idea of using relative entropy to model hyponymy is based on the assumption that the distinguishability of one word from another given its usual contexts provides us with a good metric for hyponymy. For example, if one is given a sentence with the word *dog* crossed out, it will be not be possible for sure to know whether the crossed out word is not *animal* just from the context (except perhaps very particular decelerational sentences which rely on world knowledge, such as 'All – bark'.)

Definition 8. *The **(quantum) relative entropy** of two density matrices ρ and σ is $N(\rho||\sigma) := tr(\rho \log \rho) - tr(\rho \log \sigma)$, where $0 \log 0 = 0$ and $x \log 0 = \infty$ when $x \neq 0$ by convention.*

Definition 9. *The **representativeness** between ρ and σ is $R(\rho, \sigma) := 1/(1 + N(\rho||\sigma))$, where $N(\rho||\sigma)$ is the quantum relative entropy between ρ and σ.*

Quantum relative entropy is always non-negative. For two density matrices ρ and σ, $N(\rho||\sigma) = \infty$ if $supp(\rho) \cap ker(\sigma) \neq 0$, and is finite otherwise. The following is a direct consequence of these properties:

Corollary 1. *For all density matrices ρ and σ, $R(\rho, \sigma) \leq 1$ with equality if and only if $\rho = \sigma$, and $0 \leq R(\rho, \sigma)$ with equality if and only if $supp(\rho) \cap ker(\sigma) \neq 0$*

The second part of the corollary reflects the idea that if there is a context in which it is appropriate to use v but not w, then v is perfectly distinguishable from w. Such contexts are exactly those that fall within $supp(\rho) \cap ker(\sigma)$.

Characterizing Hyponyms. The quantitative measure on density matrices given by representativeness provide a qualitative preorder on meaning representations as follows:

$$\rho \prec \sigma \text{ if } R(\rho, \sigma) > 0$$
$$\rho \sim \sigma \text{ if } \prec \sigma \text{ and } \sigma \prec \rho.$$

Proposition 1. *The following are equivalent:*

1. $\rho \prec \sigma$
2. $supp(\rho) \subseteq supp(\sigma)$
3. There exists a positive operator ρ' and $p > 0$ such that $\sigma = p\rho + \rho'$.

Proof. (1) ⇒ (2) and (2) ⇒ (1) follow directly from Corollary 1.

(2) ⇒ (3) since supp(ρ) ⊆ supp(σ) implies that there exists a $p > 0$ such that $\sigma - p\rho$ is positive. Setting $\rho' = \sigma - p\rho$ gives the desired equality.

(3) ⇒ (2) since $p > 0$, and so supp(ρ) ⊆ supp(σ) = supp($p\rho + \rho'$).

The equivalence relation $\sim$ groups any two density matrices ρ and σ with supp(ρ) = supp(σ) into the same equivalence class, thus maps the set of density matrices on a Hilbert space $\mathcal{H}$ onto the set of projections on $\mathcal{H}$. The projections are in one-to-one correspondence with the subspaces of $\mathcal{H}$ and they form an orthomodular lattice, providing a link to the logical structure of the Hilbert space [26] aims to exploit by using density matrices in IR.

Let $\widehat{w}$ and $\widehat{v}$ be density matrix representations of the words v and w. Then v is a hyponym of w in this model if $\widehat{v} \prec \widehat{w}$ and $\widehat{v} \not\sim \widehat{w}$.

Notice that even though this ordering on density matrices extracts a *yes/no* answer for the question "is v a hyponym of w?", the existence of the quantitative measure lets us to also quantify the extent to which v is a hyponym of w. This provides some flexibility in characterizing hyponymy through density matrices in practice. Instead of calling v a hyponym of w even when $R(\widehat{v}, \widehat{w})$ gets arbitrarily small, one can require the representativeness to be above a certain threshold ϵ. This modification, however, has the down side of causing the transivity of hyponymy to fail.

5   From Meanings of Words to the Meanings of Sentences Passage

As in the case for **FVect** × **P**, **CPM**(**FVect**) × **P** is a compact closed category, where the compact closed maps of **CPM**(**FVect**) and **P** lift component-wise to the product category.

Definition 10. *A **meaning space** in this new category is a pair* $(V^* \otimes V, p)$ *where* $V^* \otimes V$ *is the space in which density matrices* $v : I \to V^* \otimes V$ *of the pregroup type* p *live.*

Definition 11. *Let* $v_1 v_2 \ldots v_n$ *be a string of words, each* v_i *with a meaning space representation* $\widehat{v_i} \in (V_i^* \otimes V_i, p_i)$. *Let* $x \in P$ *be a pregroup type such that* $[p_1 p_2 \ldots p_n \leq x]$. *Then the meaning density matrix for the string is defined as:*

$$v_1 \widehat{v_2 \ldots v_n} := f(\widehat{v_1} \otimes \widehat{v_2} \otimes \ldots \otimes \widehat{v_n}) \in (W^* \otimes W, x)$$

where f *is defined to be the application of the compact closed maps obtained from the reduction* $[p_1 p_2 \ldots p_n \leq x]$ *to the composite density matrix space* $(V_1 \otimes V_1^*) \otimes (V_2^* \otimes V_2) \otimes \ldots \otimes (V_n^* \otimes V_n)$.

From a high level perspective, the reduction diagrams for **CPM**(**FVect**) × **P** look no different than the original diagrams for **FVect** × **P**, except that we depict them with thick instead of thin wires. Consider the previous example:

"John likes Mary". It has the pregroup type $n(n^r s n^l)n$, and the compact closed maps obtained from the pregroup reduction is $(\epsilon^r \otimes 1 \otimes \epsilon^l)$.

One can also depict the diagram together with the internal anatomy of the density representations in **FVect**:

The graphical reductions for compact closed categories can be applied to the diagram, establishing $(\epsilon^r \otimes 1 \otimes \epsilon^l)(\widehat{John} \otimes \widehat{likes} \otimes \widehat{Mary}) = (\widehat{John} \otimes 1 \otimes \widehat{Mary}) \circ \widehat{likes}$.

As formalised in natural logic, one expects that if the subject and object of a sentence are common nouns which are, together with the verb of the sentence, moreover, upward monotone, then if these are replaced by their hyponyms, then the meanings of the original and the modified sentences would preserve this hyponymy. The following proposition shows that the sentence meaning map for simple transitive sentences achieves exactly that.

Theorem 3. *If $\rho, \sigma, \delta, \gamma \in (N^* \otimes N, n)$, $\alpha, \beta \in (N^* \otimes N \otimes S^* \otimes S \otimes N^* \otimes N, n^l s n^r$ $\rho \prec \sigma$, $\delta \prec \gamma$ and $\alpha \prec \beta$ then*

$$f(\rho \otimes \alpha \otimes \delta) \prec f(\sigma \otimes \beta \otimes \gamma)$$

where f is the from-meanings-of-words-to-the-meaning-of-the-sentence map in Definition 11.

Proof. If $\rho \prec \sigma$, $\delta \prec \gamma$, and $\alpha \prec \beta$, then there exists a positive operator ρ' and $r > 0$ such that $\sigma = r\rho + \rho'$, a positive operator δ' and $d > 0$ such that $\gamma = d\delta + \delta'$ and a positive operator α' and $a > 0$ such that $\beta = a\alpha + \alpha'$ by Proposition 1. Then

$$
\begin{aligned}
f(\sigma \otimes \beta \otimes \gamma) &= (\epsilon^r \otimes 1 \otimes \epsilon^l)(\sigma \otimes \beta \otimes \gamma) \\
&= (\sigma \otimes 1 \otimes \gamma) \circ \beta \\
&= ((r\rho + \rho') \otimes 1 \otimes (d\delta + \delta')) \circ (a\alpha + \alpha') \\
&= (r\rho \otimes 1 \otimes d\delta) \circ (a\alpha + \alpha') + (\rho' \otimes 1 \otimes \delta') \circ (a\alpha + \alpha') \\
&= (r\rho \otimes 1 \otimes d\delta) \circ a\alpha + (r\rho \otimes 1 \otimes d\delta) \circ \alpha' + (\rho' \otimes 1 \otimes \delta') \circ (a\alpha + \alpha'), \\
f(\rho \otimes \alpha \otimes \delta) &= (\rho \otimes 1 \otimes \delta) \circ \alpha
\end{aligned}
$$

since $r, d, a \neq 0$, $\mathrm{supp}(f(\rho \otimes \alpha \otimes \delta)) \subseteq \mathrm{supp}(f(\sigma \otimes \beta \otimes \gamma))$, which by Proposition 1 proves the theorem.

6   Truth Theoretic Examples

We present several examples that demonstrate the application of the from-meanings-of-words-to-the-meaning-of-sentence map, where the initial meaning representations of words are density matrices, and explore how the hierarchy on nouns induced by their density matrix representations carry over to a hierarchy in the sentence space.

6.1   Entailment Between Nouns

Let "lions", "sloths". "plants" and "meat" have one dimensional representations in the noun space of our model:

$$\widehat{lions} = |\overrightarrow{lions}\rangle\langle\overrightarrow{lions}| \qquad\qquad \widehat{sloths} = |\overrightarrow{sloths}\rangle\langle\overrightarrow{sloths}|$$

$$\widehat{meat} = |\overrightarrow{meat}\rangle\langle\overrightarrow{meat}| \qquad\qquad \widehat{plants} = |\overrightarrow{plants}\rangle\langle\overrightarrow{plants}|$$

Let the representation of "mammals" be a mixture of one dimensional representations of individual animals:

$$\widehat{mammals} = 1/2|\overrightarrow{lions}\rangle\langle\overrightarrow{lions}| + 1/2|\overrightarrow{sloths}\rangle\langle\overrightarrow{sloths}|$$

Notice that

$$N(\widehat{lions}\|\widehat{mammals}) = tr(\widehat{lions}\log\widehat{lions}) - tr(\widehat{lions}\log\widehat{mammals})$$

$$= \log 1 - \frac{1}{2}\log\frac{1}{2} = 1$$

Hence $R(\widehat{lions}, \widehat{mammals}) = 1/2$. For the other direction, since the intersection of the support of $\widehat{mammals}$ and the kernel of $\widehat{lions}$ is non-empty, $R(\widehat{mammals}, \widehat{lions}) = 0$. This confirms that $\widehat{lions} \prec \widehat{mammals}$.

6.2   Entailment Between Sentences in One Dimensional Truth Theoretic Space

Consider a sentence space that is one dimensional, where 1 stands for true and 0 for false. Let sloths eat plants and lions eat meat; this is represented as follows

$$\widehat{eat} = (|\overrightarrow{sloths}\rangle|\overrightarrow{plants}\rangle + |\overrightarrow{lions}\rangle|\overrightarrow{meat}\rangle)(\langle\overrightarrow{sloths}|\langle\overrightarrow{plants}| + \langle\overrightarrow{lions}|\langle\overrightarrow{meat}|)$$

$$\approx (|\overrightarrow{sloths}\rangle\langle\overrightarrow{sloths}| \otimes |\overrightarrow{plants}\rangle\langle\overrightarrow{plants}|) + (|\overrightarrow{sloths}\rangle\langle\overrightarrow{lions}| \otimes |\overrightarrow{plants}\rangle\langle\overrightarrow{meat}|)$$

$$+ (|\overrightarrow{lions}\rangle\langle\overrightarrow{sloths}| \otimes |\overrightarrow{meat}\rangle\langle\overrightarrow{plants}|) + (|\overrightarrow{lions}\rangle\langle\overrightarrow{lions}| \otimes |\overrightarrow{meat}\rangle\langle\overrightarrow{meat}|)$$

The above is the density matrix representation of a pure composite state that relate "sloths" to "plants" and "lions" to "meat". If we fix the bases $\{\overrightarrow{lions}, \overrightarrow{sloths}\}$ for N_1, and $\{\overrightarrow{meat}, \overrightarrow{plants}\}$ for N_2, we will have $\widehat{eat} : N_1 \otimes N_1 \to N_2 \otimes N_2$ with the following matrix representation:

$$\begin{pmatrix} 1 & 0 & 0 & 1 \\ 0 & 0 & 0 & 0 \\ 0 & 0 & 0 & 0 \\ 1 & 0 & 0 & 1 \end{pmatrix}$$

"Lions Eat Meat". This is a transitive sentence, so as before, it has the pregroup type: $nn^l sn^r n$. Explicit calculations for its meaning give:

$$(\epsilon_N^l \otimes 1_S \otimes \epsilon_N^r)(\widehat{lions} \otimes \widehat{eat} \otimes \widehat{meat})$$

$$= \langle \overrightarrow{lions}|\overrightarrow{sloths}\rangle^2 \langle \overrightarrow{plants}|\overrightarrow{meat}\rangle^2$$

$$+ \langle \overrightarrow{lions}|\overrightarrow{sloths}\rangle \langle \overrightarrow{lions}|\overrightarrow{lions}\rangle \langle \overrightarrow{meat}|\overrightarrow{meat}\rangle \langle \overrightarrow{plants}|\overrightarrow{meat}\rangle$$

$$+ \langle \overrightarrow{lions}|\overrightarrow{lions}\rangle \langle \overrightarrow{lions}|\overrightarrow{sloths}\rangle \langle \overrightarrow{meat}|\overrightarrow{meat}\rangle \langle \overrightarrow{plants}|\overrightarrow{meat}\rangle$$

$$+ \langle \overrightarrow{lions}|\overrightarrow{lions}\rangle^2 \langle \overrightarrow{meat}|\overrightarrow{meat}\rangle^2$$

$$= 1$$

"Sloths Eat Meat". This sentence has a very similar calculation to the one above with the resulting meaning:

$$(\epsilon_N^l \otimes 1_S \otimes \epsilon_N^r)(\widehat{sloths} \otimes \widehat{eat} \otimes \widehat{meat}) = 0$$

"Mammals Eat Meat". This sentence has the following meaning calculation:

$$(\epsilon_N^l \otimes 1_S \otimes \epsilon_N^r)(\widehat{mammals} \otimes \widehat{eat} \otimes \widehat{meat}) =$$

$$(\epsilon_N^l \otimes 1_S \otimes \epsilon_N^r)((\tfrac{1}{2}\widehat{lions} + \tfrac{1}{2}\widehat{sloths}) \otimes \widehat{eat} \otimes \widehat{meat}) =$$

$$\tfrac{1}{2}(\epsilon_N^l \otimes 1_S \otimes \epsilon_N^r)(\widehat{lions} \otimes \widehat{eat} \otimes \widehat{meat}) + \tfrac{1}{2}(\epsilon_N^l \otimes 1_S \otimes \epsilon_N^r)(\widehat{sloths} \otimes \widehat{eat} \otimes \widehat{meat}) = \tfrac{1}{2}$$

The resulting meaning of this sentence is a mixture of "lions eat meat", which is true, and "sloths eat meat" which is false. Thus the value $1/2$ can be interpreted as being neither completely true or completely false: the sentence "mammals eat meat" is true for certain mammals and false for others.

6.3 Entailment Between Sentences in Two Dimensional Truth Theoretic Space

The two dimensional truth theoretic space is set as follows:

$$true \equiv |0\rangle \equiv \begin{pmatrix} 1 \\ 0 \end{pmatrix} \qquad false \equiv |1\rangle \equiv \begin{pmatrix} 0 \\ 1 \end{pmatrix}$$

The corresponding *true* and *false* density matrices are $|0\rangle\langle 0|$ and $|1\rangle\langle 1|$.

In the two dimensional space, the representation of "eats" is set as follows. Let $A = \{lions, sloths\}$ and $B = \{meat, plants\}$, then

$$\widehat{eat} \equiv \sum_{\substack{a_1,a_2 \in A \\ b_1,b_2 \in B}} |\overrightarrow{a_1}\rangle\langle\overrightarrow{a_2}| \otimes |\overrightarrow{x}\rangle\langle\overrightarrow{x}| \otimes |\overrightarrow{b_1}\rangle\langle\overrightarrow{b_2}|$$

where

$$|x\rangle \equiv \begin{cases} |0\rangle & \text{if } |a_1\rangle|b_1\rangle, |a_2\rangle|b_2\rangle \in \{|\overrightarrow{lions}\rangle|\overrightarrow{meat}\rangle, |\overrightarrow{sloths}\rangle|\overrightarrow{plants}\rangle\} \\ |1\rangle & \text{otherwise} \end{cases}$$

The generalized matrix representation of this verb in the spirit of [14] is:

$$\begin{pmatrix} 1\ 0\ 0\ 1 & 0\ 1\ 1\ 0 \\ 0\ 0\ 0\ 0 & 1\ 1\ 1\ 1 \\ 0\ 0\ 0\ 0 & 1\ 1\ 1\ 1 \\ 1\ 0\ 0\ 1 & 0\ 1\ 1\ 0 \end{pmatrix}$$

"Lions Eat Meat". The calculation for the meaning of this sentence is almost exactly the same as the case of the one dimensional meaning, only the result is not the scalar that stands for *true* but its density matrix:

$$(\epsilon_N^l \otimes 1_S \otimes \epsilon_N^r)(\widehat{lions} \otimes \widehat{eat} \otimes \widehat{meat}) = |0\rangle\langle 0|$$

"Sloths Eat Meat". Likewise, the calculation for the meaning of this sentence returns *false*:

$$(\epsilon_N^l \otimes 1_S \otimes \epsilon_N^r)(\widehat{sloths} \otimes \widehat{eat} \otimes \widehat{meat}) = |1\rangle\langle 1|$$

"Mammals Eat Meat". As we saw before, this sentence has the meaning that is the mixture of "Lions eat meat" and "Sloths eat meat"; here, this is expressed as follows:

$$(\epsilon_N^l \otimes 1_S \otimes \epsilon_N^r)(\widehat{mammals} \otimes \widehat{eat} \otimes \widehat{meat})$$
$$= \frac{1}{2}(\epsilon_N^l \otimes 1_S \otimes \epsilon_N^r)(\widehat{lions} \otimes \widehat{eat} \otimes \widehat{meat}) + \frac{1}{2}(\epsilon_N^l \otimes 1_S \otimes \epsilon_N^r)(\widehat{sloths} \otimes \widehat{eat} \otimes \widehat{meat})$$
$$= \frac{1}{2}|1\rangle\langle 1| + \frac{1}{2}|0\rangle\langle 0|$$

So in a two dimensional truth theoretic model, "Mammals eat meat" give the completely mixed state in the sentence space, which has maximal entropy. This is equivalent to saying that we have no real knowledge whether mammals in general eat meat or not. Even if we are completely certain about whether individual mammals that span our space for "mammals" eat meat, this information differs uniformly within the members of the class, so we cannot generalize.

Already with a two dimensional truth theoretic model, the relation $\widehat{lions} \prec \widehat{mammals}$ carries over to sentences. To see this, first note that we have

$$N(\widehat{lions\ eat\ meat}||\widehat{mammals\ eat\ meat}) = N\left(|0\rangle\langle 0| \ \middle\| \ \frac{1}{2}|0\rangle\langle 0| + \frac{1}{2}|1\rangle\langle 1|\right)$$

$$= (|0\rangle\langle 0|)\log(|0\rangle\langle 0|) - (|0\rangle\langle 0|)\log\left(\frac{1}{2}|0\rangle\langle 0| + \frac{1}{2}|1\rangle\langle 1|\right) = 1$$

In the other direction, we have $N(\widehat{mammals\,eat\,meat} \| \widehat{lions\,eat\,meat}) = \infty$, since the intersection of the support of the first argument and the kernel of the second argument is non-trivial. These lead to the following representativeness results between sentences:

$$R(\widehat{lions\,eat\,meat}, \widehat{mammals\,eat\,meat}) = 1/2$$

$$R(\widehat{mammals\,eat\,meat} \| \widehat{lions\,eat\,meat}) = 0$$

As a result we obtain:

$$\widehat{lions\,eat\,meat} \prec \widehat{mammals\,eat\,meat}$$

Since these two sentences share the same verb phrase, from-meaning-of-words-to-the-meaning-of-sentence map carries the hyponymy relation in the subject words of the respective sentences to the resulting sentence meanings. By using the density matrix representations of word meanings together with the categorical map from the meanings of words to the meanings of sentences, the knowledge that a lion is an animal lets us infer that "mammals eat meat" implies "lions eat meat":

$$(\widehat{lions} \prec \widehat{mammals}) \rightarrow (\widehat{lions\,eat\,meat} \prec \widehat{mammals\,eat\,meat})$$

"Dogs Eat Meat". To see how the completely mixed state differs from a perfectly correlated but pure state in the context of linguistic meaning, consider a new noun $\overrightarrow{dog} = |\overrightarrow{dog}\rangle\langle\overrightarrow{dog}|$ and redefine eat in terms of the bases $\{\overrightarrow{lions}, \overrightarrow{dogs}\}$ and $\{\overrightarrow{meat}, \overrightarrow{plants}\}$, so that it will reflect the fact that dogs eat itboth meat and plants. We define "eat" so that it results in the value of being "half-true half-false" when it takes "dogs" as subject and "meat" or "plants" as object. The value "half-true half-false" is the superposition of *true* and *false*: $\frac{1}{2}|0\rangle + \frac{1}{2}|1\rangle$. With this assumptions, $\widehat{eat}$ will still be a pure state with the following representation in **FVect**:

$$|\overrightarrow{eat}\rangle = |\overrightarrow{lions}\rangle \otimes |0\rangle \otimes |\overrightarrow{meat}\rangle + |\overrightarrow{lions}\rangle \otimes |1\rangle \otimes |\overrightarrow{plants}\rangle$$

$$+ |\overrightarrow{dogs}\rangle \otimes (\frac{1}{2}|0\rangle + \frac{1}{2}|1\rangle) \otimes |\overrightarrow{meat}\rangle + |\overrightarrow{dogs}\rangle \otimes (\frac{1}{2}|0\rangle + \frac{1}{2}|1\rangle) \otimes |\overrightarrow{plants}\rangle$$

Hence, the density matrix representation of "eat" becomes:

$$\widehat{eat} = |\overrightarrow{eat}\rangle\langle\overrightarrow{eat}|$$

The calculation for the meaning of the sentence is as follows:

$$(\epsilon_N^l \otimes 1_S \otimes \epsilon_N^r)(\widehat{dogs} \otimes \widehat{eat} \otimes \widehat{meat})$$

$$= (\epsilon_N^l \otimes 1_S \otimes \epsilon_N^r)(|\overrightarrow{dogs}\rangle\langle\overrightarrow{dogs}| \otimes |\overrightarrow{eat}\rangle\langle\overrightarrow{eat}| \otimes |\overrightarrow{meat}\rangle\langle\overrightarrow{meat}|)$$

$$= (\frac{1}{2}|0\rangle + \frac{1}{2}|1\rangle)(\frac{1}{2}\langle0| + \frac{1}{2}\langle1|)$$

So in this case, we are certain that it is half-true and half-false that dogs eat meat. This is in contrast with the completely mixed state we got from "Mammals eat meat", for which the truth or falsity of the sentence was entirely unknown.

"Mammals eat meat", again. Let "mammals" now be defined as:

$$\widehat{mammals} = \frac{1}{2}\widehat{lions} + \frac{1}{2}\widehat{dogs}$$

The calculation for the meaning of this sentence gives:

$$(\epsilon_N^l \otimes 1_S \otimes \epsilon_N^r)(\widehat{mammals} \otimes \widehat{eat} \otimes \widehat{meat})$$

$$= \frac{1}{2}(\epsilon_N^l \otimes 1_S \otimes \epsilon_N^r)(\widehat{lions} \otimes \widehat{eat} \otimes \widehat{meat}) + \frac{1}{2}(\epsilon_N^l \otimes 1_S \otimes \epsilon_N^r)(\widehat{dogs} \otimes \widehat{eat} \otimes \widehat{meat})$$

$$= \frac{3}{4}|0\rangle\langle 0| + \frac{1}{4}|0\rangle\langle 1| + \frac{1}{4}|1\rangle\langle 0| + \frac{1}{4}|1\rangle\langle 1|$$

This time the resulting sentence representation is not completely mixed. This means that we can generalize the knowledge we have from the specific instances of mammals to the entire class to some extent, but still we cannot generalize completely. This is a mixed state, which indicates that even if the sentence is closer to *true* than to *false*, the degree of truth isn't homogeneous throughout the elements of the class. The non-zero non-diagonals indicate that it is also partially correlated, which means that there are some instances of "mammals" for which this sentence is true to a degree, but not completely. The relative similarity measures between *true* and *false* and the sentence can be calculated explicitly using fidelity:

$$F\big(|1\rangle\langle 1|, \widehat{mammals\,eat\,meat}\big) = \langle 1|\widehat{mammals\,eat\,meat}|1\rangle = \frac{1}{4}$$

$$F\big(|0\rangle\langle 0|, \widehat{mammals\,eat\,meat}\big) = \langle 0|\widehat{mammals\,eat\,meat}|0\rangle = \frac{3}{4}$$

Notice that these values are different from the values for the representativeness for truth and falsity of the sentence, even thought they are proportional: the more representative their density matrices, the more similar the sentences are to each other. For example, we have:

$$N\big(|1\rangle\langle 1| \parallel \widehat{mammals\,eat\,meat}\big) =$$

$$\mathrm{tr}\big(|1\rangle\langle 1|\big)\log(|1\rangle\langle 1|)\big) - \mathrm{tr}\big(|1\rangle\langle 1|\log(\frac{3}{4}|0\rangle\langle 0| + \frac{1}{4}|0\rangle\langle 1| + \frac{1}{4}|1\rangle\langle 0| + \frac{1}{4}|1\rangle\langle 1|)\big) \approx 2$$

Hence, $R\big(|1\rangle\langle 1| \parallel \widehat{mammals\,eat\,meat}\big) \approx .33$. On the other hand:

$$N\big(|0\rangle\langle 0| \parallel \widehat{mammals\,eat\,meat}\big) =$$

$$\mathrm{tr}\big(|0\rangle\langle 0|\big)\log(|0\rangle\langle 0|)\big) - \mathrm{tr}\big(|0\rangle\langle 0|\log(\frac{3}{4}|0\rangle\langle 0| + \frac{1}{4}|0\rangle\langle 1| + \frac{1}{4}|1\rangle\langle 0| + \frac{1}{4}|1\rangle\langle 1|)\big) \approx 0.41$$

Hence, $R\big(|0\rangle\langle 0| \parallel \widehat{mammals\,eat\,meat}\big) \approx 0.71$.

7   Distributional Examples

The goal of this section is to show how one can obtain density matrices for words using lexical taxonomies and co-occurrence frequencies counted from corpora of text. We show how these density matrices are used in example sentences and how the density matrices of their meanings look like. We compute the representativeness formula for these sentences to provide a proof of concept that this measure does makes sense for data harvested from corpora distributionally and that its application is not restricted to truth-theoretic models. Implementing these constructions on real data and validating them on large scale datasets constitute work in progress.

7.1   Entailment Between Nouns

Suppose we have a noun space N. Let the subspace relevant for this part of the example be spanned by lemmas *pub, pitcher, tonic*. Assume that the (non-normalized version of the) vectors of the atomic words *lager* and *ale* in this subspace are as follows:

$$\overrightarrow{lager} = 6 \times \overrightarrow{pub} + 5 \times \overrightarrow{pitcher} + 0 \times \overrightarrow{tonic}$$
$$\overrightarrow{ale} = 7 \times \overrightarrow{pub} + 3 \times \overrightarrow{pitcher} + 0 \times \overrightarrow{tonic}$$

Suppose further that we are given taxonomies such as 'beer = lager + ale', harvested from a resource such as WordNet. Atomic words (i.e. leafs of the taxonomy), correspond to *pure* states and their density matrices are the projections onto the one dimensional subspace spanned by $|\overrightarrow{w}\rangle\langle\overrightarrow{w}|$. Non-atomic words (such as *beer*) are also density matrices, harvested from the corpus using a feature-based method similar to that of [12]. This is done by counting (and normalising) the frequency of times a word has co-occurred with a subset B of bases in a window in which other bases (the ones not in B) have not occurred.

Formally, for a subset of bases $\{b1, b2, ..., bn\}$, we collect co-ordinates C_{ij} for each tuple $|bi\rangle|bj\rangle$ and build the density matrix $\sum_{ij} C_{ij}|bi\rangle|bj\rangle$.

For example, suppose we see *beer* six times with just *pub*, seven times with both *pub* and *pitcher*, and none-whatsoever with *tonic*. Its corresponding density matrix will be as follows:

$$\widehat{beer} = 6 \times |\overrightarrow{pub}\rangle\langle\overrightarrow{pub}| + 7 \times (|\overrightarrow{pub}\rangle + |\overrightarrow{pitcher}\rangle)(\langle\overrightarrow{pub}| + \langle\overrightarrow{pitcher}|)$$

$$= 13 \times |\overrightarrow{pub}\rangle\langle\overrightarrow{pub}| + 7 \times |\overrightarrow{pub}\rangle\langle\overrightarrow{pitcher}| + 7 \times |\overrightarrow{pitcher}\rangle\langle\overrightarrow{pub}| + 7 \times |\overrightarrow{pitcher}\rangle\langle\overrightarrow{pitcher}|$$

To calculate the similarity and representativeness of the word pairs, we first normalize them via the operation $\frac{\rho}{\text{Tr}\rho}$, then apply the corresponding formulae. For example, the degree of similarity between 'beer' and 'lager' using fidelity is as follows:

$$\text{Tr}\sqrt{\widehat{lager}^{\frac{1}{2}} \cdot \widehat{beer} \cdot \widehat{lager}^{\frac{1}{2}}} = 0.93$$

The degree of entailment $lager \prec beer$ is 0.82 as computed as follows:

$$\frac{1}{1 + \text{Tr}(\widehat{lager} \cdot \log(\widehat{lager}) - \widehat{lager} \cdot \log(\widehat{beer}))} = 0.82$$

The degree of entailment $beer \prec lager$ is 0, like one would expect.

7.2   Entailment Between Sentences

To see how the entailment between sentences follows from the entailment between words, consider example sentences *'Psychiatrist is drinking lager'* and *'Doctor is drinking beer'*. For the sake of brevity, we assume the meanings of *psychiatrist* and *doctor* are mixtures of basis elements, as follows:

$$\widehat{psychiatrist} = 2 \times |\overrightarrow{patient}\rangle\langle\overrightarrow{patient}| + 5 \times |\overrightarrow{mental}\rangle\langle\overrightarrow{mental}|$$

$$\widehat{doctor} = 5 \times |\overrightarrow{patient}\rangle\langle\overrightarrow{patient}| + 2 \times |\overrightarrow{mental}\rangle\langle\overrightarrow{mental}| + 3 \times |\overrightarrow{surgery}\rangle\langle\overrightarrow{surgery}|$$

The similarity between *psychiatrist* and *doctor* is:

$$S(\widehat{psychiatrist}, \widehat{doctor}) = S(\widehat{doctor}, \widehat{psychiatrist}) = 0.76$$

The representativeness between them is:

$$R(\widehat{psychiatrist}, \widehat{doctor}) = 0.49 \qquad R(\widehat{doctor}, \widehat{psychiatrist}) = 0$$

We build matrices for the verb *drink* following the method of [15]. Intuitively this is as follows: the value in entry (i, j) of this matrix will reflect how typical it is for the verb to have a subject related to the ith basis and an object related to the jth basis. We assume that the small part of the matrix that interests us for this example is as follows:

Drink	Pub	Pitcher	Tonic
Patient	4	5	3
Mental	6	3	2
Surgery	1	2	1

This representation can be seen as a pure state living in a second order tensor. Therefore the density matrix representation of the same object is $\widehat{drink} = |\overrightarrow{drink}\rangle\langle\overrightarrow{drink}|$, a fourth order tensor. Lifting the simplifications introduced in [15] from vectors to density matrices, we obtain the following linear algebraic closed forms for the meaning of the sentences:

$$\widehat{Psychiatrist\ is\ drinking\ lager} = \widehat{drink} \odot (\widehat{psychiatrist} \otimes \widehat{lager})$$

$$\widehat{Doctor\ is\ drinking\ beer} = \widehat{drink} \odot (\widehat{doctor} \otimes \widehat{beer})$$

Applying the fidelity and representativeness formulae to sentence representations, we obtain the following values:

$$S(\widehat{Psychiatrist\ is\ drinking}\ lager, \widehat{Doctor\ is\ drinking}\ beer) = 0.81$$

$$R(\widehat{Psychiatrist\ is\ drinking}\ lager, \widehat{Doctor\ is\ drinking}\ beer) = 0.53$$

$$R(\widehat{Doctor\ is\ drinking}\ beer, \widehat{Psychiatrist\ is\ drinking}\ lager) = 0$$

From the relations psychiatrist $\prec$ doctor and lager $\prec$ beer we obtain the desired entailment between sentences:

$$Psychiatrist\ is\ drinking\ lager \prec Doctor\ is\ drinking\ beer.$$

The entailment between these two sentences follows from the entailment between their subjects and the entailment between their objects. In the examples that we have considered so far, the verbs of sentences are the same. This is not a necessity. One can have entailment between sentences that do not have the same verbs, but where the verbs entail each other, examples can be found in [2]. The reason we do not present such cases here is lack of space.

8 Conclusion and Future Work

The often stated long term goal of compositional distributional models is to merge distributional and formal semantics. However, what formal and distributional semantics *do* with the resulting meaning representations is quite different. Distributional semanticists care about *similarity* while formal semanticists aim to capture *truth* and *inference*. In this work we presented a theory of meaning using basic objects that will not confine us to the realm of only distributional or only formal semantics. The immediate next step is to develop methods for obtaining density matrix representations of words from corpus, that are more robust to statistical noise, and testing the usefulness of the theory in large scale experiments.

The problem of integrating function words such as 'and', 'or', 'not', 'every' into a distributional setting has been notoriously hard. We hope that the characterization of compositional distributional entailment on these very simple types of sentences will provide a foundation on which we can define representations of these function words, and develop a more logical theory of compositional distributional meaning.

References

1. Abramsky, S., Coecke, B.: A categorical semantics of quantum protocols. In: Proceedings of the 19th Annual IEEE Symposium on Logic in Computer Science, pp. 415–425. IEEE Computer Science Press (2004). arXiv:quant-ph/0402130
2. Balkır, E.: Using density matrices in a compositional distributional model of meaning. Master's thesis, University of Oxford (2014)

3. Baroni, M., Bernardi, R., Do, N-Q., Shan, C-C.: Entailment above the word level in distributional semantics. In: Proceedings of the 13th Conference of the European Chapter of the Association for Computational Linguistics, pp. 23–32. Association of Computational Linguists (2012)
4. Beltagy, I., Chau, C., Boleda, G., Garrette, D., Erk, K., Mooney, R.: Montague meets markov: deep semantics with probabilistic logical form. In: Second Joint Conference on Lexical and Computational Semantics, vol. 1, pp. 11–21. Association of Computational Linguists (2013)
5. Blacoe, W., Kashefi, E., Lapata, M.: A quantum-theoretic approach to distributional semantics. In: Proceedings of the 2013 Conference of the North American Chapter of the Association for Computational Linguistics: Human Language Technologies, pp. 847–857 (2013)
6. Bruza, P.D., Cole, R.: Quantum logic of semantic space: an exploratory investigation of context effects in practical reasoning. In: We Will Show Them: Essays in Honour of Dov Gabbay, pp. 339–361 (2005)
7. Clark, S., Coecke, B., Sadrzadeh, M.: A compositional distributional model of meaning. In: Proceedings of the Second Symposium on Quantum Interaction (QI-2008), pp. 133–140 (2008)
8. Clarke, D.: Context-theoretic semantics for natural language: an overview. In: Proceedings of the Workshop on Geometrical Models of Natural Language Semantics, pp. 112–119. Association for Computational Linguistics (2009)
9. Coecke, B.: Quantum picturalism. Contemp. Phys. **51**(1), 59–83 (2010)
10. Coecke, B., Sadrzadeh, M., Clark, S.: Mathematical foundations for a compositional distributional model of meaning. Linguist. Anal. **36** (2010)
11. Firth, John R.: A Synopsis of Linguistic Theory, 1930–1955. Studies in Linguistic, Analysis, pp. 1–32 (1957)
12. Geffet, M., Dagan, I.: The distributional inclusion hypotheses and lexical entailment. In: Proceedings of the 43rd Annual Meeting on Association for Computational Linguistics, pp. 107–114. Association for Computational Linguistics (2005)
13. Glickman, O., Dagan, I., Koppel, M.: Web based probabilistic textual entailment. In: Proceedings of the PASCAL Challenges Workshop on Recognizing Textual Entailment (2005)
14. Grefenstette, E.: Category-Theoretic Quantitative Compositional Distributional Models of Natural Language Semantics. Ph.D. thesis, University of Oxford (2013)
15. Grefenstette, E., Sadrzadeh, M.: Experimental support for a categorical compositional distributional model of meaning. In: Proceedings of the Conference on Empirical Methods in Natural Language Processing, pp. 1394–1404. Association for Computational Linguistics (2011)
16. Herbelot, A., Ganesalingam, M.: Measuring semantic content in distributional vectors. In: Proceedings of the 51st Annual Meeting of the Association for Computational Linguistics, vol. 2, pp. 440–445. Association for Computational Linguistics (2013)
17. Kotlerman, L., Dagan, I., Szpektor, I., Zhitomirsky-Geffet, M.: Directional distributional similarity for lexical inference. Nat. Lang. Eng. **16**(04), 359–389 (2010)
18. Lambek, J.: Type grammars as pregroups. Grammars **4**(1), 21–39 (2001)
19. Lenci, A., Benotto, G.: Identifying hypernyms in distributional semantic spaces. In: Proceedings of the First Joint Conference on Lexical and Computational Semantics, vol. 2, pp. 75–79. Association for Computational Linguistics (2012)
20. Lin, D.: An information-theoretic definition of similarity. In: Proceedings of the International Conference on Machine Learning, pp. 296–304 (1998)

21. MacCartney, B., Manning, C.D.: Natural logic for textual inference. In: ACL Workshop on Textual Entailment and Paraphrasing, Association for Computational Linguistics (2007)
22. Piedeleu, R., Kartsaklis, D., Coecke, B., Sadrzadeh, M.: Open system categorical quantum semantics in natural language processing (2015). arXiv:1502.00831
23. Santus, E., Lenci, A., Lu, Q., Walde, S.S.I.: Chasing hypernyms in vector spaces with entropy. In: Proceedings of the 14th Conference of the European Chapter of the Association for Computational Linguistics, vol. 2, pp. 38–42 (2014)
24. Selinger, P.: Dagger compact closed categories and completely positive maps. Electron. Notes Theoret. Comput. Sci. **170**, 139–163 (2007)
25. Sordoni, A., Nie, J-Y., Bengio, Y.: Modeling term dependencies with quantum language models for ir. In: Proceedings of the 36th International ACM SIGIR Conference on Research and Development in Information Retrieval, pp. 653–662. Association for Computational Linguistics (2013)
26. Van Rijsbergen, C.J.: The Geometry of Information Retrieval. Cambridge University Press, New York (2004)
27. Weeds, J., Weir, D., McCarthy, D.: Characterising measures of lexical distributional similarity. In: Proceedings of the 20th International Conference on Computational Linguistics, Number 1015. Association for Computational Linguistics (2004)

On Symmetric and Choiceless Computation

Anuj Dawar[(⊠)]

University of Cambridge Computer Laboratory, William Gates Building,
J.J. Thomson Avenue, Cambridge CB3 0FD, UK
anuj.dawar@cl.cam.ac.uk

Formal models of computation such as Turing machines are usually defined as performing operations on strings of symbols. Indeed, for most purposes, it suffices to consider strings over a two-letter alphabet $\{0, 1\}$. Decision problems are defined as sets of strings, and complexity classes as sets of decision problems. However, many natural algorithms are described on more abstract structure (such as graphs) because this is the natural level of abstraction at which to describe the problem being solved. Of course, we know that the abstract structures can be ultimately represented as strings (and, indeed, have to be in actual computational devices), but the representation comes at a cost. The same abstract structure may have many different string representations and the implementation of the algorithm may break the intended abstraction.

Research in the area of finite model theory and descriptive complexity (see [11,13]) has, over the years, developed a number of techniques of describing algorithms and complexity classes directly on classes of relational structures, rather than strings. Along with this, many methods of proving inexpressiblity results have been shown, often described in terms of games. A key question that has been the focus of this research effort is whether the complexity class P admits a descriptive characterisation (see [10, Chap. 11]).

A recent paper [1] ties some of the logics studied in finite model theory to natural circuit complexity classes, and shows thereby that inexpressibility results obtained in finite model theory can be understood as lower bound results on such classes. In this presentation, I develop the methods for proving lower bound results in the form of combinatorial arguments on circuits, without reference to logical definability. The present abstract gives a brief account of the results and methods.

Symmetric Circuits. We start with a brief introduction to the formalism of circuit complexity. A language $L \subseteq \{0, 1\}^*$ can be described as a family of *Boolean functions*: $(f_n)_{n \in \omega} : \{0, 1\}^n \to \{0, 1\}$. Each f_n can be represented by a *circuit* C_n which is a directed acyclic graph where we think of the vertices as gates suitably labeled by Boolean operators $\wedge, \vee, \neg$ for the internal gates and by inputs $x_1, \ldots, x_n$ for the gates without incoming edges. One gate is distinguished as determining the output. If there is a polynomial $p(n)$ bounding the size of C_n (i.e. the number of gates in C_n), then the language L is said to be in the complexity class P/poly. If, in addition, the family of circuits is *uniform*, meaning that the function that takes n to C_n is itself computable in polynomial time, then L is in P. For the definition of either of these classes, it does not make a

© IFIP International Federation for Information Processing 2016
Published by Springer International Publishing Switzerland 2016. All Rights Reserved.
M.T. Hajiaghayi and M.R. Mousavi (Eds.): TTCS 2015, LNCS 9541, pp. 23–29, 2016.
DOI: 10.1007/978-3-319-28678-5_2

difference if we expand the class of gates that we can use in the circuit beyond the Boolean basis to include, for instance, *threshold* or *majority* gates. The presence of such gates can make a difference for more restricted circuit complexity classes, for instance when we limit the depth of the circuit to be bounded by a constant, but not when we allow arbitrary polynomial-size circuits. Also, in the circuit characterization of P, it does not make a difference if we replace the uniformity condition with a stronger requirement. Say, we might require that the function taking n to C_n is computable in DLogTime.

We are interested in languages that represent properties of relational structures such as graphs. For simplicity, let us restrict attention to directed graphs, i.e. structures in a vocabulary with one binary relation. A property of such graphs that is in P can be recognised by a family $(C_n)_{n \in \omega}$ of Boolean circuits of polynomial size and uniformity, as before, where now the inputs to C_n are labelled by the n^2 *potential edges* of an n-vertex graph, each taking a value of 0 or 1. Given an n-vertex graph G, there are many ways that it can be mapped onto the inputs of the circuit C_n, one for each bijection between $V(G)$—the vertices of G—and $[n]$. So, to ensure that the family of circuits is really defining a property of graphs, we require it to be *invariant* under the choice of this mapping. That is, each input of C_n carries a label of the form (i, j) for $i, j \in [n]$ and we require the output to be unchanged under any permutation $\pi \in S_n$ acting on the inputs by the action $(i, j) \mapsto (\pi(i), \pi(j))$. It is clear that any property of graphs that is invariant under isomorphisms of graphs and is in P is decided by such a family of circuits. Say that a circuit C_n is *symmetric* if any permutation $\pi \in S_n$ can be extended to an *automorphism* of C_n which takes each input (i, j) to $(\pi(i), \pi(j))$. Below, we do not distinguish notationally between the permutation π and its extension to an automorphism of C_n.

Lower Bounds for Symmetric Circuits. It is clear that symmetric circuits are necessarily invariant and it is not difficult to come up with examples that show that the converse is not true. That is, we can show that there are polynomial-time decidable properties of graphs that are not decided by polynomial-size families of Boolean circuits. Indeed, one can show that the property of a graph G having an even number of edges cannot be decided by any such family. It is rather more challenging to show that there are polynomial time decidable properties that are not decided by polynomial-size families of symmetric circuits with threshold gates. The proof is based on three ingredients, which are elaborated next.

Support Theorem. The first is a combinatorial analysis of symmetric circuits establishing the so-called *bounded support property*. For a gate g in C_n, a symmetric circuit taking n-vertex graphs as input, we say that a set $X \subseteq [n]$ *supports* g if for every $\pi \in S_n$ such that $\pi(x) = x$ for all $x \in X$, we also have $\pi(g) = g$. The *support theorem* in [1] establishes that if $(C_n)_{n \in \omega}$ is a family of symmetric circuits of polynomial size then there is a k such that all gates in C_n have a support of at most k elements.

Theorem 1 [1]. *For any polynomial p, there is a k such that for all sufficiently large n, if C is a symmetric circuit on $[n]$ of size at most $p(n)$, then every gate in C has a support of size at most k.*

The theorem as proved in [1] establishes this in greater generality, allowing for circuit sizes that group super-polynomially and yielding support sizes that are then also non-constant. However, the simpler version suffices for the lower bounds state here. The proof is a combinatorial analysis of the action of permutations on a symmetric circuit and the interested reader should refer to [1] for details.

Bijection Games. The second ingredient in the lower bound proof is the bijection game of Hella [12], which in combination with the support theorem, gives us a tool for showing that certain pairs of graphs are not distinguished by any circuit where all gates have bounded size supports. Hella defined this game as a characterization of equivalence in a fragment of first-order logic where the number of variables is limited, but we have *counting quantifiers*. The logic does not concern us here. We regard the game as defining a family of equivalence relations $\equiv^k$, parameterized by a positive integer k. The game is played on graphs G and H (or, more generally, finite relational structures) by two players called Spoiler and Duplicator using pebbles $a_1, \ldots, a_k$ on G and $b_1, \ldots, b_k$ on H. At any point in the game, the pebbles may be placed on vertices of the respective graphs and we do not distinguish notationally below between the pebble and the vertex on which it is placed. One move of the game proceeds as follows:

- Spoiler chooses a pair of pebbles a_i and b_i;
- Duplicator chooses a bijection $h : V(G) \to V(H)$ such that for pebbles a_j and b_j $(j \neq i)$, $h(a_j) = b_j$; and
- Spoiler chooses $a \in V(G)$ and places a_i on a and b_i on $h(a)$.

If, after this move, the map $a_1 \ldots a_k \mapsto b_1 \ldots b_k$ is not an isomorphism between the subgraphs of G and H induced by the pebbled vertices, the game is over and Spoiler wins, otherwise it can continue. We say that $G \equiv^k H$ if Duplicator has a strategy for playing forever. Clearly, if G and H are isomorphic, then $G \equiv^k H$ as Duplicator can always play the isomorphism as its choice of bijection. Conversely, if $k \geq n$ then Spoiler can force a win as long as G and H are not isomorphic. Thus $G \equiv^n H$ implies that G and H are isomorphic. For smaller values of k, the equivalence relation provides an approximation of ismorphism. The family of equivalence relations so defined is also known as the Weisfeiler-Lehman equivalences (see [5] for an account).

What links these games with symmetric circuits is the following.

Theorem 2. *If C is a symmetric circuit on n-vertex graphs such that every gate of C has a support of size at most k, and G and H are graphs such that $G \equiv^{2k} H$ then, C accepts G if, and only if, C accepts H.*

Proof. (Sketch). To prove this, we show that if C distinguishes G from H, it provides a winning strategy for Spoiler in the $2k$-pebble bijection game played on G and H, which guarantees a win in at most kd moves, where d is the depth of the circuit C. Specifically, Spoiler plays by maintaining a pointer to a gate g of C and a bijection $\alpha : V(G) \to [n]$ between the vertices of G and $[n]$ so that the following conditions are satisfied in any game position $(\bar{u}, \bar{v})$ that arises in the game:

– $\alpha(\bar{u})$ includes the support of g; and
– for any bijection $\beta : V(H) \to [n]$ such that $\beta^{-1}\alpha(\bar{u}) = \bar{v}$, we have $C_g(\alpha(G)) \neq C_g(\beta(H))$.

Here, $C_g(\alpha(G))$ denotes the value that the gate g takes in the evaluation of the circuit C when the inputs are assigned the edges of G according to the map α. Similarly, $C_g(\beta(H))$ denotes the value that the gate g takes in the evaluation of the circuit C when the inputs are assigned the edges of H according to the map β.

These conditions are initially satisfied by taking α to be any bijection and letting g be the output gate of C, since C is assumed to be symmetric and to distinguish G from H. The key step in the proof shows that, given that the conditions are satisifed, Spoiler can, within k moves, move the pointer to a child of g so that the conditions are still satisfied. Indeed, suppose $\gamma : V(G) \to V(H)$ is the bijection that Duplicator plays. Since, by assumption, g is evaluated differently under the assignments $\alpha(G)$ and $\alpha\gamma^{-1}(H)$, there must be a child h of g so that $C_h(\alpha(G)) \neq C_h(\alpha\gamma^{-1}(H))$. Spoiler aims to place pebbles on the vertices of G in $\alpha^{-1}(s)$ where s is the support of h, within k moves. At each move, Duplicator may change the bijection γ. However, since the elements corresponding to the support of g are pebbled, there is an automorphism of C that fixes g, corresponding to the changed bijection. This enables us to show that Spoiler can indeed force pebbles onto the elements $\alpha^{-1}(s)$ where s is the support of a suitable h, within k moves, while maintaining the conditions (1) and (2) above.

Thus, within kd moves, the conditions are satisfied with the pointer at a gate which is an input gate of C, say labelled with the input (i, j). The support of this gate is just the set $\{i, j\}$ so, by assumption, there are pebbles on $u = \alpha^{-1}(i)$ and $v = \alpha^{-1}(j)$ in G and corresponding pebbles on $\gamma(u)$ and $\gamma(v)$. The condition that $C_g(\alpha(G)) \neq C_g(\alpha\gamma^{-1}(H))$ tells us that one of these is an edge and the other is not, which means that Spoiler has won the bijection game.

Cai-Fürer Immerman Graphs. The final ingredient in proving a lower bound is the construction of pairs of graphs that are not isomorphic, but are equivalent in the relation $\equiv^k$. Just such a construction is provided by Cai et al. [5]. To be precise, they show that there is sequence of pairs of graphs G_k and H_k ($k \in \mathbb{N}$) so that for each k we have $G_k \equiv^k H_k$ and there is a polynomial-time decidable property of graphs that includes all the G_k and excludes all H_k. In particular, it follows from our discussion above that this polynomial-time property is not decided by any polynomial-size family of *symmetric* circuits, even in the presence of threshold gates.

The graphs G_k and H_k are obtained from a single graph G by replacing the vertices and edges of G by suitably defined gadgets. In particular, each edge of

G is replaced in G_k and H_k by a pair of parallel edges. Swapping the endpoints of one such pair of edges distinguishes G_k from H_k. It can then be shown that if the graph G we start from is sufficiently well connected, in particular if it has tree-width at least k, then $G_k \equiv^k H_k$. For details of the construction, including the connection to tree-width and pebble games, we refer the reader to [9].

Reductions and Complete Problems. Having established a super-polynomial lower bound for symmetric threshold circuits for one (artificial) problem, we are able to tranfer such lower bounds to other problems by means of reductions. The appropriate notion of reduction here is a symmetric version of AC^0 reductions, i.e. reductions given by families of *constant-depth, polynomial-size* Boolean circuits. In order to formally define such reductions, we have to consider circuits which do not have a single output but can, instead, be used to define a relation on $[n]$ when presented with an n-vertex graph, and we need to extend the definition of symmetry to such circuits. This is done formally in [1]. To be precise, we have a circuit C along with an injective function $\Omega : [n] \to C$. The requirement for symmetry now says any permutation $\pi \in S_n$ extends to an automorphism of C so that $\pi(\Omega(x)) = \Omega(\pi(x))$.

Reductions defined by constant-depth symmetric circuits are closely related to reductions given by formulas of first-order logic (FO-reductions). In particular, it is easy to show that if a problem P is recucible to Q by means of an FO-reduction, it is reducible by means of a symmetric AC^0-reduction. It is known from the work of Lovász and Gács [14] that a version of the Boolean satisfiability problem SAT (suitably represented as a class of relational structures) is NP-complete under FO-reductions. It follows immediately that this problem cannot be solved by polynomial-size symmetric threshold circuits. If it could, all problems in NP would be solved by such circuit families and we have already established that this is not the case. A similar result holds for the class of Hamiltonian graphs, since this is also known to be NP-complete under FO-reductions by a construction due to Dahlhaus [6].

On the other hand, there are natural graph problems, such as 3-colourability, which are NP-complete under the usual sense of polynomial-time reductions but which are provably not NP-complete under FO-reductions (the latter follows from results in [8]). It remains an open question whether 3-SAT (with a natural representation as a class of relational structures) is NP-complete under FO-reductions. Nonetheless, it is possible to use FO-reductions from established lower bounds to show that neither 3-SAT nor 3-colourability is solvable by polynomial-size symmetric threshold circuits (these results follow from constructions in [3]). In the case of 3-colourability, there is also a construction that deploys bijection games (as in [7]) directly to show that this problem is not decidable by such families. These lower bounds should be contrasted with the fact that the existence of perfect matchings in graphs is definable in the logic FPC [2] and therefore is decidable by polynomial-size symmetric threshold circuits.

The following table gives a list of problems that are known to be solvable by polynomial-size symmetric threshold circuits (under *upper bounds*) and a list of closely related problems that are provably not solvable by such families (under *lower bounds*). It should be noted that the lower bound results are *unconditional* in that they do not rely on unproved complexity-theoretic assumptions such as $P \neq NP$.

Upper bounds	Lower bounds
Circuit value problem	SAT
2-Colourability	3-Colourability
2-SAT	3-SAT
Perfect matching	Hamiltonian cycle
Linear programming	XOR-SAT
Isomorphism on planar graphs	Isomorphism on bounded-degree graphs

Choiceless Computation. Another way of describing algorithms that work directly with relational structures rather than strings that encode them are the *abstract state machines* of Gurevich (see [4] and references therein). Here, the ability of a classical machine to make arbitrary choices is eschewed in favour of a high degree of parallelism. While such machines are universal, and can describe any algorithm at a high level of abstraction, it remains an open question whether every polynomial-time decidable class of structures is decidable in polynomial-time by such a machine. Specifically, Blass et al. [4] define the class CPTC of problems decidable in *choiceless polynomial time with counting* and ask the question whether this includes all of P. They conjecture that it does not, but the question remains unresolved. An interesting angle to this question is to examine it through the lens of circuits.

References

1. Anderson, M., Dawar, A.: On symmetric circuits and fixed-point logics. In: 31st International Symposium Theoretical Aspects of Computer Science (STACS 2014), pp. 41–52 (2014)
2. Anderson, M., Dawar, A., Holm, B.: Maximum matching and linear programming in fixed-point logic with counting. In: 28th Annual ACM/IEEE Symposium Logic in Computer Science, pp. 173–182 (2013)
3. Atserias, A., Bulatov, A., Dawar, A.: Affine systems of equations and counting infinitary logic. Theor. Comput. Sci. **410**(18), 1666–1683 (2009)
4. Blass, A., Gurevich, Y., Shelah, S.: Choiceless polynomial time. Ann. Pure Appl. Logic **100**, 141–187 (1999)
5. Cai, J.-Y., Fürer, M., Immerman, N.: An optimal lower bound on the number of variables for graph identification. Combinatorica **12**(4), 389–410 (1992)
6. Dahlhaus, E.: Reduction to NP-Complete Problems by Interpretation. LNCS, vol. 171. Springer, London (1984)
7. Dawar, A.: A restricted second order logic for finite structures. Inf. Comput. **143**, 154–174 (1998)

8. Dawar, A., Grädel, E.: Properties of almost all graphs and generalized quantifiers. Fundamenta Informaticae **98**(4), 351–372 (2010)
9. Dawar, A., Richerby, D.: The power of counting logics on restricted classes of finite structures. In: Duparc, J., Henzinger, T.A. (eds.) CSL 2007. LNCS, vol. 4646, pp. 84–98. Springer, Heidelberg (2007)
10. Ebbinghaus, H.-D., Flum, J.: Finite Model Theory, 2nd edn. Springer, Heidelberg (1999)
11. Grädel, E., Kolaitis, P.G., Libkin, L., Marx, M., Spencer, J., Vardi, M.Y., Venema, Y., Weinstein, S.: Finite Model Theory and Its Applications. Springer, Heidelberg (2007)
12. Hella, L.: Logical hierarchies in PTIME. In: Proceedings of the 7th IEEE Symposium on Logic in Computer Science, pp. 360–368 (1992)
13. Immerman, N.: Descriptive Complexity. Springer, New York (1999)
14. Lovász, L., Gács, P.: Some remarks on generalized spectra. Zeitschrift für Mathematische Logik und Grundlagen der Mathematik **23**, 27–144 (1977)

Robots' Cooperation for Finding a Target in Streets

Mohammad Abouei Mehrizi[1]([✉]), Mohammad Ghodsi[2],
and Azadeh Tabatabaei[1]

[1] Department of Computer Engineering, Sharif University of Technology,
Tehran, Iran
{abouei,atabatabaei}@ce.sharif.edu
[2] School of Computer Science Institute for Research in Fundamental Sciences (IPM),
Sharif University of Technology, Tehran, Iran
ghodsi@sharif.edu

Abstract. We study the problem of finding a target t from a start point s in street environments with the cooperation of two robots which have a minimal sensing capability; that is, robots do not know any information about the workspace including information on distances, edges, coordinates, angles etc. They just can detect the discontinuities in the visibility region of their location. The robots can detect target point t as soon as it enters their visibility region and have communication peripherals to send messages to each other. Our aim is to minimize the length of the path passed by the robots. We propose an online algorithm for robots such that they move in the workspace and find the target. This algorithm generates a search path from a start point s to a target point t such that the distance traveled by the robots is at most 2 times longer than the shortest path. Also, we prove that this ratio is tight.

Keywords: Computational geometry · Visibility · Motion planning · Minimal sensing · Multi robot's cooperation

1 Introduction

The problem of finding a target in an environment is a fundamental problem in computational geometry and robotics [5,7,10]. This problem is known as exploration which appears in many applications where the environment is unknown and no geometric map of the scene is available. In this problem, the robot's sensors are the only input device to gather information. Depending on the ability of the sensors, information will be exact. Simple sensor robots have many benefits such as low cost, less sensitivity to failure, stable against noise etc. [3].

In this research robots have a limited ability in sensing the environment. They have an abstract sensor to detect discontinuities of the work space (called gaps) in their visibility region. Discontinuity means a portion of the environment that is not visible to the robot (Fig. 1). The robot can assigning a label L or R to

© IFIP International Federation for Information Processing 2016
Published by Springer International Publishing Switzerland 2016. All Rights Reserved.
M.T. Hajiaghayi and M.R. Mousavi (Eds.): TTCS 2015, LNCS 9541, pp. 30–43, 2016.
DOI: 10.1007/978-3-319-28678-5_3

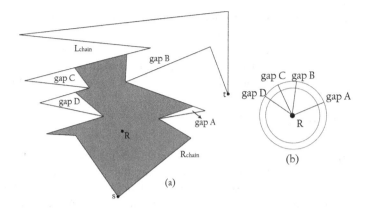

Fig. 1. (a) A street in which L_{chain} is the left chain and R_{chain} is the right chain. The colored region is the visibility polygon of the point robot R in the street. (b) The position of discontinuities in the depth information detected by the sensor. Note that robot cannot detect the angular information of gaps, it just can detect the position of gap and move toward them (Color figure online).

every gap g. When robot is scanning the environment counter clockwise, if a gap is sensed such that discontinuity is near to far then gap is a right gap (labeled by R), otherwise it is a left gap (labeled by L). In Fig. 1 (a) A and B are right gaps while C and D are left gaps. Also, the robots are capable of recognizing a target point t when it is in the robots' visibility region.

Robots do not need any data structure to store the topological map of the environment and just store some fixed data for navigation. They have to move towards the gaps to cover the hidden region behind each gap. Note that the robots do not have to measure any angles or distances to the walls of the scene or infer their position but have an infinite and omnidirectional field of view. Robots' cooperation will be done by sending messages to each other. Cooperation is necessary for some situation. For example, when one robot find out its path is along the shortest path (from s to t), it send a message and help the other robot to find the correct path. Messages are some kind of data known for both robots. In this paper, the workspace is assumed to be a restricted simple polygon called a street.

Definition 1. *A simple polygon P with two vertices s and t is called a street if the counter-clockwise polygonal chain R_{chain} from s to t and the clockwise chain L_{chain} from s to t are mutually weakly visible* [6].

Definition 1 means each point on the left chain L_{chain} must be visible from at least one point on the right chain R_{chain} and vice versa (Fig. 1). A street polygon is also known as L-R visible polygon [2].

This paper will present an algorithm such that two point robots that are equipped with a gap sensor and communication device start navigating from s to reach target t. The robots have no geometric map of the scene and have to make decisions to achieve the target only based on the information gathered through the sensor and received messages.

This paper is organized as follows. Section 2 will review related works, Sect. 3 will present structures and the model used; Sect. 4 will present preliminaries. Section 5 will present the algorithm, and Sect. 6 will present the conclusions.

2 Related Works

Because of application and usefulness of path finding in online problems, it has recently received much attention from researchers. Klein proposed the first competitive online strategy for searching a target point in a street [6]; The robot employed in [6] is equipped with a 360 degree vision system. Also, it can measure each angle or distance to the walls of the street. As the robot moves, a partial map is constructed from what has been seen so far. Klein proved an upper bound of 5.72 for the competitive ratio (the ratio of the length of the traversed path to the shortest path from s to t) of this problem. Also, it was proved later that there is no strategy with the competitive ratio less than $\sqrt{2}$ for this problem.

A strategy similar to Klein's with the competitive ratio of $\pi+1$ has been introduced in [8,9] which is robust under small navigation errors. Other researchers have presented several algorithms with the competitive ratios between $\sqrt{2}$ and the upper bound of 5.72 [7,9]. Icking et al. presented an optimal strategy with the competitive ratio of $\sqrt{2}$ [5].

The limited sensing model that we use in this paper was first introduced by Lavalle et al. [13]. Gap Navigation Tree (GNT) has been proposed to maintain and update the gaps seen along the navigating path. This tree is built by detecting the discontinuities in the depth information and updated by the topological changes of the information. The topological changes are appearances, disappearances, merges, and splits of gaps.

Another minimal sensing model was introduced by Suri and Vicari [11] for a simple robot. They assume that the robot can only sense the combinatorial (non-metric) properties of their surroundings. The sensor can detect vertices of the polygon in its visibility region, and can report if there is a polygon edge between consecutive vertices. The information is maintained in two combinatorial vectors, called the combinatorial visibility vector (cvv) and the point identification vector (piv). Despite minimal capability, they show that the robot can obtain enormous geometric reasoning and can accomplish many non-trivial tasks.

Tabatabaei and Ghodsi [23] presented an online algorithm to find the target point t from the start point s with one minimal sensing robot in the street environment. Beacuse their model is minimal sensing their problem is more difficult than Kleins problem. Their strategy is doubling to find the target such that the robot will move 11 times as long as the shortest path with just one pebble and if the robot has access to many pebbles, this ratio reduces to 9. In the doubling strategy, the robot moves back and forth on the line, such that at each stage i, it walks 2^i steps in one direction, comes back to the origin, walks 2^{i+1} steps in the opposite direction until the target is reached. Also pebble is some thing portable and distinguishable by the robot. Showing that this ratio is optimal in this model, they presented an optimal strategy for walking in streets with minimum number of turns [14].

Path finding for multiple robots where paths are collision-free is still an interesting and difficult problem which has been studied for many years [19–22]. Jingjin Yu and Steven M. LaValle reported a reduction from a special case of multi agent path planning to network flow [18]. They introduced a multiflow based ILP algorithm for planning optimal, collision-free paths for multiple robots on graphs, introducing complete ILP algorithms to solve time optimal and distance optimal multi-robot path planning problems [16]. They also proved that computing a minimum total arrival time, a minimum makespan, or a minimum total distance solution is NP-complete for some kinds of cooperation multi-robot path planning [17].

Luna and Bekris proposed an algorithm that guarantees completeness for a general class of problems without any assumptions about the graph's topology. Their algorithm required two primitive operations: push? where agents move towards their goals and do this to arrive a point where no progress can be made. Swap allows two agents to swap their positions. Their approach can address any solvable instance in a graph of size n where there are at most $n-2$ agents [20].

Mikael Hammar et al. [15] have presented constant competitive strategies to explore a rectilinear simple polygon in the L_1 metric with one or more robots. They also proved there is no deterministic strategy for path exploration of a rectilinear polygon with one robot that has competitive ratio $2 - \epsilon$ for any $\epsilon > 0$. They also showed there are no deterministic strategies for exploring a rectilinear polygon with two or three robots having smaller competitive ratio than $\frac{2}{3}$ but their results are not tight.

In this paper we propose an online search strategy for two point robots equipped with the gap sensor and a communication peripheral like Bluetooth, Wi-Fi with wireless antenna, etc. with respect to expansion of environment. The aim of the algorithm is to minimize the minimum path length of the two robots from the start point s to the target point t in a street workspace. We will show the minimum search path which is generated by the strategy is at most 2 times as long as the shortest path that is tight for this situation. To our knowledge, this is the first result providing some competitive ratio for two robots walking in streets in a minimal sensing model.

3   The Sensing Model and Communications

3.1   Gap Sensor

Gap sensor used in this paper is a visual sensing model. At any position q of the environment, a cyclically ordered location of the depth discontinuities in the visibility region of the point is what the robot's sensor detects, as shown in (Fig. 1). When the robot reports the discontinuities counterclockwise from a visibility region, it assigns a left label to a transition from far to near and assigns a right label to a transition from near to far [12] (Fig. 2). The robot can only

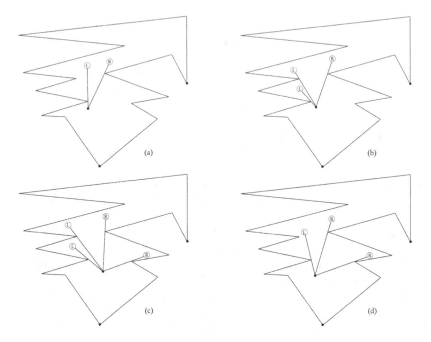

Fig. 2. The dark circle denotes the location of the robot. (a) Existing gaps at the beginning. (b) A split event. (c) An appearance event. (d) A merge event.

walk towards the gaps, this restriction cannot get us into trouble and despite this constraint we will present a 2-approximation algorithm.

GNT data structure has been introduced as a means for the robot system to navigate in an unknown scene [12]. This structure is bassed on four critical events. In the strategy proposed here, the robots do not need to store GNT; they just pay attention to the events which will change the number of gaps. Each robot just stores two special gaps (later will discuss about these gaps) and sets its orientation to one of them and moves toward it until a critical event occurs.

3.2   Communications

In the algorithm (presented in Sect. 5) we will use two robots to find the target. To achieve a good ratio, robots are equipped with communication peripherals for sending data together. The type of communication device depends on whether the expansion of workspace can be detected. For example, if robots will not have more than a 100 m distance, we can use Bluetooth as the communication device. If more than 100 m is needed, we can use Wi-Fi devices and wireless network antenna prepared in the workspace.

In this research we will not explain communication methods and their potential difficulties, like message failure. We assume that robots can have communication together and send messages to each other as well as possible.

3.3   Motion Primitive

Robots in their movement can sense gaps and their labels (right or left) in counter-clockwise order. Gaps' labels are not unique and just show that the gap is a left gap (L) or a right one (R) (Fig. 2). These gaps may change during the robots' movement. Robots can set their direction to a gap and move toward it. During walking they cannot detect geometric properties of the environment. They can only walk to gaps until a critical event occurs or the gap disappears; then they will make a decision and find a direction for movement and walk around the workspace. Also, robots can detect and move toward target t as soon as it enters the visibility region.

4   Preliminaries

In this paper we use the notation used in [14]; some of the definitions and lemmas that will be used in our strategy are mentioned here.

At each point p of the robots' search paths, their sensor either sees the target or achieves a set of gaps with the label of L or R (*l-gap* and *r-gap* for abbreviation). If the target is visible, the robot moves towards the target to reach it. When the robots report the position of the gaps, they should move towards the gaps to find the target.

Definition 2. *In the set of l-gaps, the gap that is on the right side of the others is called most-advanced left gap and is denoted by g_l. Analogously, in the set of r-gaps, the gap that is on the left side of the others is called most-advanced right gap and is denoted by g_r* [14].

Note: Accourding to the Definition 2, we know in visibility region of robots, all of the right gaps are near together and then all of the left gaps are near to each other (counter clockwise). Definition 2 means when robot is scanning its visibility region counter clockwise, the last right gap is called most-advanced right gap and the first left gap is called most-advanced left gap. In any street polygon (circular, general etc.) this is correct.

Lemma 1. *At any point of the robot's search paths, if the target is not visible, then it is behind one of the most-advanced gaps* [14].

As the robots move in the environment, g_l and g_r may dynamically change. The critical events in which the structure of the robot's visibility region changes can also change g_l and g_r. In the next section, we show how the critical events change the left most advanced gaps such that a sequence of the left most advanced gaps, $[g_{l1}, g_{l2}, ..., g_{lm}]$, appears in the robots' visibility region, while exploring the street. Similarly, a sequence of the right most advanced gaps, $[g_{r1}, g_{r2}, ..., g_{rn}]$, may occur.

At each point, if there is exactly one of the two gaps (g_l or g_r), then the goal is hidden behind that gap. Thus, there is no ambiguity and the robots move toward the gap. If both g_r and g_l exist, then the target is hidden behind one of these gaps. This case is called a *funnel*. As soon as a robot enters a point in which both g_r and g_l exist, a funnel situation starts. This case continues until one of g_l or g_r disappears or until they become collinear. When a robot enters a point in which there is a funnel situation, it will wait for the other robot. In this situation the only non-trivial case in this navigation occurs. Before describing our strategy, we state some features of a street and the gaps that are applied in the algorithm.

When the robot enters a funnel situation, there are two convex chains in front of it: the left convex chain that lies on the left chain (L_{chain}) of the street, and the right convex chain that lies on the right chain (R_{chain}) of the street. The two chains have the following main property.

Lemma 2. *When a funnel situation starts, the shortest path from s to t lies completely on the left convex chain, or on the right convex chain of the funnel* [14].

Theorem 1. *For any vertex $v_j \in L_{chain}$ (or, $v_j \in R_{chain}$), the shortest path from s to v_j makes a left turn (respectively, a right turn) at every vertex of L_{chain} (respectively, R_{chain}) in the path* [4].

Lemma 3. *Each of the two convex chains, in a funnel situation, contains a point at which the funnel situation ends or a new funnel situation starts* [14].

Note: Lemmas 1, 2 and 3 are about the features of environment (street polygon) and are not related to the robots[1].

5   Algorithm

In this section we explain our movement strategy for the robots to move in the environment (street polygon) from s to t. We use R_1 for the first robot and R_2 for the second one. Note that both robots start at point s, and if R_1(respectively, R_2) achieves t, it will send a message (*target-founded*) to the R_2(respectively, R_1); when R_2(respectively, R_1) receives the message, it stops searching. This moment represents the end of movement for both robots.

Obviously, this strategy minimizes the robots' search paths. That is, our aim is:

$$Minimize : Min\{path_1, path_2\}$$

Such that $path_1$ is the length of the path traversed by R_1 and $path_2$ is the length of the path traversed by R_2 from s to t.

As mentioned, since our sensing model is minimal, then robots decide to move in the environment based on the information gathered by their sensor. The intuition behind the algorithm is when robots are in the funnel situation, then according to Lemma 1, when the target is visible from neither R_1 nor R_2,

---

[1] Capabilities or number of robots does not have any relation to these lemmas.

then the target is behind one of the most-advanced gaps. Because robots do not know which gap is correct, then R_1 will move to g_l and R_2 will move to g_r; since we refer to R_1 as R_{left} and R_2 as R_{right}. If there is just one of g_l or g_r, for each robot, it will get moving toward that gap.

In the set of l-gaps, the gap which is on the right side of the others for R_{left} is called *the most left robot advanced left gap* and is denoted by g_{ll}; In the set of r-gaps, the gap which is on the left side of the others for R_{left} is called *the most left robot advanced right gap* and is denoted by g_{lr}. Analogously, In the set of l-gaps, the gap which is on the right side of the others for R_{right} is called *the most right robot advanced left gap* and is denoted by g_{rl}; in the set of r-gaps, the gap which is on the left side of the others for R_{right} is called *the most right robot advanced right gap* and is denoted by g_{rr} (Fig. 3).

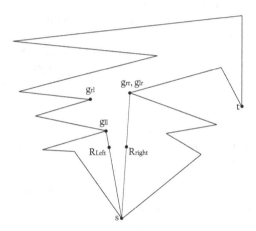

Fig. 3. The red line is the direction of R_{right} and blue line is the direction of R_{left} from the start point s. R_{right} moves to g_{rr} and R_{left} moves to g_{ll}. In the visibility region of R_{right}, gaps are different from those in the visibility region of R_{left} (Color figure online).

We know in each funnel situation either R_{left} or R_{right} passes the shortest path. If there is just one advanced gap (g_l or g_r), the target is behind it, and then both robots move toward it. If a robot finds out that its path is along the shortest path, it will send a message to the other robot and the other will correct its direction.

5.1 Critical Events

In any funnel situation, both robots start from one point and search the workspace. During the movement, some critical events may occur for each robot. Critical events usually occur when the visibility region of a robot changes such that it makes a change to the direction of robots. These events may happen for either R_{left} or R_{right} and change their direction. Events that may occur while the robots are moving and the corresponding operations are mentioned bellow.

i. When the R_{right} crosses a bitangent complement of g_{rl} and an r-gap, then g_{rr} will be replaced by the r-gap (Fig. 4.a).
ii. When the R_{left} crosses a bitangent complement of g_{lr} and an l-gap, then g_{ll} will be replaced by the l-gap (Fig. 4.b).
iii. When the R_{right} crosses a bitangent complement of g_{rl} and another l-gap, then g_{rl} splits and will be replaced by the l-gap (Fig. 4.c).
iv. When the R_{left} crosses a bitangent complement of g_{lr} and another r-gap, then g_{lr} splits and will be replaced by the r-gap (Fig. 4.d).
v. When the R_{right} crosses a bitangent of g_{rr} and another r-gap, at the point in which g_{rr} disappears, g_{rr} will be replaced by the r-gap (disappearance and split events occur simultaneously). In this situation, if there are more than one gap similar to the r-gap, g_{rr} will be replaced by the one which is on the left side of the others (Fig. 4.e).
vi. When the R_{left} crosses a bitangent of g_{ll} and another l-gap, at the point in which g_{ll} disappears, g_{ll} will be replaced by the l-gap (disappearance and split events occur simultaneously). In this situation, if there are more than one gap similar to the l-gap, g_{ll} will be replaced by the one which is on the right side of the others (Fig. 4.f).
vii. When the R_{left}/R_{right} crosses over an inflection ray, each of g_{ll}/g_{rl} or g_{rl}/g_{rr} which is adjacent to the ray, disappears and is eliminated from the detected gaps automatically.

If R_{right} goes to g_{rr} and event i occurs there will be two different situations (let $r_{right} \in r$-gap be the right gap that event i occured for it):

- $\| R_{right} - g_{rl} \| < \| R_{right} - r_{right} \|$. This situation is shown in Fig. 4 (a), after replacing g_{rr}, R_{right} will change its direction to g_{rl} (g_{rl} and g_{rr} are collinear) and keep going until it touches it. When R_{right} touches g_{rl}, there may occur some critical events that update g_{rr} or/and g_{rl}. Then R_{right} will continue its routing by going toward g_{rr}.
- $\| R_{right} - g_{rl} \| > \| R_{right} - r_{right} \|$. In this situation R_{right} is shown in Fig. 5 after replacing the g_{rr}, like before, R_{right} will change its direction to g_{rl} (g_{rl} and g_{rr} are collinear) and keep going until it touches it. But when R_{right} goes to g_{rl} there will occur two different events.

  • g_{rr} disappears and there is no split event (no new r-gap (g_{rr}) will be created): in this situation R_{right} moves toward to g_{rl} until an split event occurs and g_{rr} is found (Fig. 5 (a)).
  • When the robot arrives at g_{rr}, event v happens and g_{rr} will update. In this situation R_{right} will move toward g_{rl} until touches it. Then R_{right} will keep going to the g_{rr} (Fig. 5 (b)).
  Note: In Fig. 5 (b) just two gaps are visible at start point s, because robots are capable of detecting the discontinuities of environment and other reflex vertices do not create any gap, then robots cannot distinguish them.

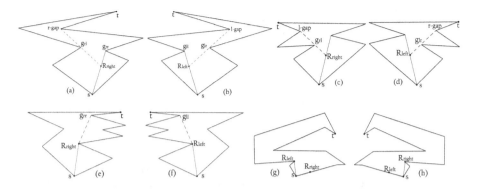

Fig. 4. Critical events will occur during the robots movement. (a) R_{right} finds a new *r-gap* in its visibility region which is on left side of g_{rr}. (b) R_{left} finds a new *l-gap* in its visibility region which is on right side of g_{ll}. (c) Split event occurs for *l-gaps* of R_{right}. (d) Split event occurs for *r-gaps* of R_{left}. (e) Split event occurs for *r-gaps* of R_{right}. (f) Split event occurs for *l-gaps* of R_{left}. (g) Critical message event occurs for R_{right}; when R_{left} arrives at point g_{ll}, g_{ll} disappears and a new right gap will appear. It finds out that its path is along the shortest path and sends *bad-direction* message to R_{right}, and R_{right} corrects its direction. (h) Like (g) critical message event occurs for R_{left}.

When R_{right} is going to g_{rr} no critical event occur (appearance a new gap is not a critical event) until R_{right} crosses bitangent complement of g_{rl} and an *r-gap* (event i), then the above situation will be done.

When R_{left} goes to g_{ll} and event ii occurs, R_{left} will act like R_{right} as mentioned above.

Note: When R_{right} crosses a bitangent complement of g_{rl} and an *r-gap*, it knows that it has to change its direction to g_{rl} and it does not need to measure any distance; It just change its direction to g_{rl} and keep going, during movement some critical events will happen and R_{right} can update its important gaps (g_{rl}, g_{rr}).

5.2   Message Events

These events occur when a robot receives a message from the other, then it will manipulate it and do corresponding operations. These message events are mention here.

1. When the R_{left} crosses a bitangent of g_{ll} and an *r-gap*, at the point in which g_{ll} disappears and the *r-gap* appears due to split event (disappearance and split events occur simultaneously), R_{left} sends a bad-direction message to R_{right}, then R_{right} will change its direction to g_{rl} (Fig. 4.g).
2. When the R_{right} crosses a bitangent of g_{rr} and an l-gap, at the point in which g_{rr} disappears and the *l-gap* appears, R_{right} sends a *bad-direction* message to R_{left}, then R_{left} will change its direction to g_{lr} (Fig. 4.h).

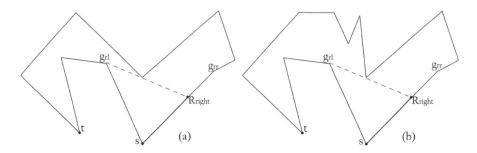

Fig. 5. R_{right} crosses the bitangent of g_{rl} and an $r\text{-}gap$ such that $\| R_{right} - r\text{-}gap \| < \| R_{right} - g_{rl} \|$. (a) When R_{right} goes to the g_{rl} there will not be any right gap. (b) When R_{right} goes to g_{rl} there will be right gap and g_{rr} is found.

Both robots will keep going until the current funnel situation ends and arrives at a point which is the start of next funnel situation. Then the earlier robot will send a *funnel-situation* message to another one and waits there until the other robot arrives. When a robot receives a *funnel-situation* message it will keep going until it arrives at the new funnel situation, then it sends a *start-funnel* to the other robot and both do this algorithm for new funnels until finding the target t.

From the start point of any funnel situation, R_{left} just saves g_{ll}, g_{lr} and moves toward g_{ll}; and R_{right} just saves g_{rr}, g_{rl} and moves toward g_{rr}. Note that in the start point s, the gaps coincide for both robots, but when they move in the environment, their gaps may be different from each other (Fig. 3). R_{left} will keep going to g_{ll} and R_{right} will keep going to g_{rr} all of the time, as often as each of the following events occurs (robots will ignore the other events):

– When one of the critical events that are mentioned in Sect. 5.1 occurs.
– When one of the message events that are mentioned in Sect. 5.2 occurs.

Note: Some other events may occur during robots' movement, but they are not important for us. It means, robots just pay attention to mentioned events in Sects. 5.1 and 5.2, robots will ignore the other events.

5.3   Analysis

Lemma 4. *In the strategy mentioned in Sect. 5, robot R_{right}/R_{left} will change its direction only when the robot crosses bitangent of g_{rl}/g_{lr} and an $r\text{-}gap/l\text{-}gap$, or receives a message event.*

Proof. Directly concludes from the critical and message events presented in Sects. 5.1 and 5.2.

Lemma 5. *Assume both robots are in point s and are in the funnel situation, R_{left} moves toward g_{ll} and R_{right} moves toward g_{rr}, if R_{left}/R_{right} passes along the shortest path to arrive at g_{ll}/g_{rr}, then R_{right}/R_{left} will pass a detour with the length of at most $3 \| s - g_{ll}/g_{rr} \|$ to arrive at g_{ll}/g_{rr}.*

Proof. According to Lemma 2 either R_{left} or R_{right} passes along the shortest path. Without loss of generality, assume R_{left} passes along the shortest path and R_{right} goes in a bad direction. Let $a = \parallel s - g_{ll} \parallel$, we know R_{left} after traversing a distance of length a, understands that it has passed the shortest path and sends *bad-direction* message to R_{right}, R_{right} passes the distance a (like R_{left}), in the worst case R_{right} has to come back to the point s and then moves to g_{ll}. Then it will pass at most $3 \parallel s - g_{ll} \parallel$. This is true when R_{right} passes the shortest path and R_{left} goes in a bad direction.

Both robots will pass a path with at most 4 times as long as the shortest path because in any funnel situation one robot is along the shortest path and the other will pass a detour at most 3 times as long as the shortest path (according to Lemma 5).

Theorem 2. $Min\{path_1, path_2\} \le 2path_{optimal}$, *such that* R_{left} *passes along* $path_1$, R_{right} *passes along* $path_2$ *and* $path_{optimal}$ *is shortest path from* s *to* t.

Proof. Let α be the length of the shortest path from s to t. We know in any funnel situation one robot is along the shortest path. To maximize the length of $Min\{path_1, path_2\}$ we should take $\mid path_1 \mid = \mid path_2 \mid$ and R_{left} passes at most $\frac{\alpha}{2}$ along the shortest path and detours $\frac{3\alpha}{2}$. Then R_{left} passes at most $\frac{\alpha}{2} + \frac{3\alpha}{2} = 2\alpha$. This is the case for R_{right} too; in Fig. 6, a tight example for $2 - \epsilon(\epsilon > 0)$ ratio is shown.

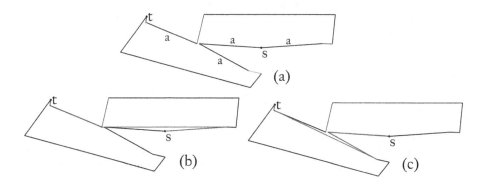

Fig. 6. (a) The workspace is shown, the pink segment on the R_{chain} is visible just from the pink point on the L_{chain} and the green segment on the L_{chain} is visible just from the green point on the R_{chain}. The shortest path from s to t has a length of $2a + \epsilon_0$. (b) The red line shows the path of R_{right} according to the presented algorithm. As we can see, R_{right} will pass a path with almost $4a + \epsilon_0 - \epsilon_1$ length. (c) The red line shows R_{left} path according to the presented algorithm. As we can see, R_{left} will pass a path with almost $4a + \epsilon_0 - \epsilon_2$ length too. Then the algorithm is tight for ratio $2 - \epsilon(\epsilon > 0)$ (Color figure online).

6 Conclusion

In this research, we proposed an online algorithm for two simple robots, such that they find the target t from the start point s. Minimal sensing means that robots do not have any sense from the environment and can only detect the discontinuities (known as gap), and move to their directions. Also, the robots can detect the target t, as soon as it enters their visibility region. We prepared an algorithm to find target such that there exists one robot whose path from s to t is at most 2 times as long as the shortest path; this ratio is tight. Introducing more general classes of polygons which admit competitive searching with minimal sensing multi robots is an interesting open problem.

References

1. Baezayates, R.A., Culberson, J.C., Rawlins, G.J.: Searching in the plane. Inf. Comput. **106**(2), 234–252 (1993)
2. Das, G., Heffernan, P.J., Narasimhan, G.: LR-visibility in polygons. Comput. Geom. **7**(1), 37–57 (1997)
3. Gfeller, B., Mihalák, M., Suri, S., Vicari, E., Widmayer, P.: Counting targets with mobile sensors in an unknown environment. In: Kutyłowski, M., Cichoń, J., Kubiak, P. (eds.) ALGOSENSORS 2007. LNCS, vol. 4837, pp. 32–45. Springer, Heidelberg (2008)
4. Ghosh, S.K.: Visibility Algorithms in the Plane. Cambridge University Press, Cambridge (2007)
5. Icking, C., Klein, R., Langetepe, E.: An optimal competitive strategy for walking in streets. In: Meinel, C., Tison, S. (eds.) STACS 1999. LNCS, vol. 1563, pp. 110–120. Springer, Heidelberg (1999)
6. Klein, R.: Walking an unknown street with bounded detour. In: Proceedings of the 32nd Annual Symposium on Foundations of Computer Science, 1991, pp. 304–313, October 1991
7. Kleinbergt, J.M.: On-line search in a simple polygon. In: Proceedings of the 5th Annual ACM-SIAM Symposium on Discrete Algorithms, No. 70, p. 8. SIAM (1994)
8. Lopez-Ortiz, A.: On-line target searching in bounded and unbounded domains. University of Waterloo (1996)
9. Lopez-Ortiz, A., Schuierer, S.: Simple, efficient and robust strategies to traverse streets. In: Proceedings of the 7th Canada Conference on Computational Geometry (1995)
10. Mitchell, J.S.: Geometric shortest paths and network optimization. Handb. Comput. Geom. **334**, 633–702 (2000)
11. Suri, S., Vicari, E., Widmayer, P.: Simple robots with minimal sensing: from local visibility to global geometry. Int. J. Robot. Res. **27**(9), 1055–1067 (2008)
12. Tovar, B., Murrieta-Cid, R., LaValle, S.M.: Distance-optimal navigation in an unknown environment without sensing distances. IEEE Trans. Robotics **23**(3), 506–518 (2007)
13. Tovar, B., LaValle, S.M., Murrieta, R.: Optimal navigation and object finding without geometric maps or localization. In: Proceedings of the IEEE International Conference on Robotics and Automation 2003, ICRA 2003, vol. 1, pp. 464–470, September 2003

14. Tabatabaei, A., Ghodsi, M.: Optimal strategy for walking in streets with minimum number of turns for a simple robot. In: Zhang, Z., Wu, L., Xu, W., Du, D.-Z. (eds.) COCOA 2014. LNCS, vol. 8881, pp. 101–112. Springer, Heidelberg (2014)
15. Hammar, M., Nilsson, B.J., Persson, M.: Competitive exploration of rectilinear polygons. In: Lingas, A., Nilsson, B.J. (eds.) FCT 2003. LNCS, vol. 2751, pp. 234–245. Springer, Heidelberg (2003)
16. Yu, J., LaValle, S.M.: Planning optimal paths for multiple robots on graphs. In: 2013 IEEE International Conference on Robotics and Automation (ICRA), pp. 3612–3617, May 2013
17. Yu, J., LaValle, S.M.: Structure and intractability of optimal multi-robot path planning on graphs. In: AAAI, June 2013
18. Yu, J., LaValle, S.M.: Multi-agent path planning and network flow. In: Frazzoli, E., Lozano-Perez, T., Roy, N., Rus, D. (eds.) Algorithmic Foundations of Robotics X. STAR, vol. 86, pp. 157–173. Springer, Heidelberg (2013)
19. Erdmann, M., Lozano-Perez, T.: On multiple moving objects. Algorithmica **2**(1–4), 477–521 (1987)
20. Luna, R., Bekris, K.E.: Push and swap: fast cooperative path-finding with completeness guarantees. In: IJCAI, pp. 294–300, July 2011
21. Ryan, M.R.K.: Exploiting subgraph structure in multi-robot path planning. J. Artif. Intell. Res. **31**, 497–542 (2008)
22. Silver, D.: Cooperative pathfinding. In: AIIDE, pp. 117–122, June 2005
23. Tabatabaei, A., Ghodsi, M.: Walking in streets with minimal sensing. J. Comb. Optim. (2014). http://link.springer.com/article/10.1007/s10878-014-9791-4

Some Properties of Continuous Yao Graph

Davood Bakhshesh and Mohammad Farshi[✉]

Combinatorial and Geometric Algorithms Laboratory, Department of Computer
Science, Yazd University, Yazd, Iran
dbakhshesh@gmail.com, mfarshi@yazd.ac.ir

Abstract. Given a set S of points in the plane and an angle $0 < \theta \leq 2\pi$,
the *continuous Yao graph* $cY(\theta)$ with vertex set S and angle θ defined
as follows. For each $p, q \in S$, we add an edge from p to q in $cY(\theta)$ if
there exists a cone with apex p and angular diameter θ such that q is
the closest point to p inside this cone.

In this paper, we prove that for $0 < \theta < \pi/3$ and $t \geq \frac{1}{1-2\sin(\theta/2)}$,
the continuous Yao graph $cY(\theta)$ is a $\mathcal{C}$-fault-tolerant geometric t-spanner
where $\mathcal{C}$ is the family of convex regions in the plane. Moreover, we show
that for every $\theta \leq \pi$ and every half-plane h, $cY(\theta) \ominus h$ is connected,
where $cY(\theta) \ominus h$ is the graph after removing all edges and points inside
h from the graph $cY(\theta)$. Also, we show that there is a set of n points in
the plane and a convex region C such that for every $\theta \geq \frac{\pi}{3}$, $cY(\theta) \ominus C$
is not connected.

Given a geometric network G and two vertices x and y of G, we call
a path P from x to y a self-approaching path, if for any point q on P,
when a point p moves continuously along the path from x to q, it always
get closer to q. A geometric graph G is self-approaching, if for every pair
of vertices x and y there exists a self-approaching path in G from x to y.
In this paper, we show that there is a set P of n points in the plane such
that for some angles θ, Yao graph on P with parameter θ is not a self-
approaching graph. Instead, the corresponding continuous Yao graph on
P is a self-approaching graph. Furthermore, in general, we show that for
every $\theta > 0$, $cY(\theta)$ is not necessarily a self-approaching graph.

Keywords: t-spanner · Region-fault tolerant spanner · Continuous Yao
graph · Self-approaching graph

1 Introduction

Let S be a set of n points in $\mathbb{R}^d$ and let $t \geq 1$ be a real number. A geometric
graph is an edge-weighted graph on $S \subseteq \mathbb{R}^d$ such that the weight of each edge is
the Euclidean distance between its endpoints. A geometric graph G with vertex
set S is called a t-spanner for S, if for each two points p and q in S, there
exists a path Q in G between p and q whose length is at most t times $|pq|$, the
Euclidean distance between p and q. The length of a path is defined to be the
sum of the lengths, or weight, of all edges on the path. The path Q is called a

© IFIP International Federation for Information Processing 2016
Published by Springer International Publishing Switzerland 2016. All Rights Reserved.
M.T. Hajiaghayi and M.R. Mousavi (Eds.): TTCS 2015, LNCS 9541, pp. 44–55, 2016.
DOI: 10.1007/978-3-319-28678-5_4

t-spanner path (or *t-path*) between p and q. We denote the length of path Q by $|Q|$. The *stretch factor* (or *dilation*) of G is the smallest value of t for which G is a t-spanner. The t-spanners were introduced by Peleg and Schäffer [14] in the scope of distributed computing and, then, by Chew [6] in the scope of computational geometry. The t-spanners are applicable in many scopes such as graph theory, network topology design, distributed systems, robotics. We refer the reader to [5,8,12,16] for reading about the t-spanners and their applications.

The problem of efficient construction of a t-spanner for a given point set and a constant $t > 1$ has been studied extensively. One can see the major algorithms for building spanners in the book by Narasimhan and Smid [13].

The Yao graph used by Andrew Yao to construct Euclidean minimum spanning tree on high-dimensional Euclidean space [17]. For a set S of points in the plane, the Yao graph Y_k, for $k \geq 2$, is defined as follows. At each point $u \in S$, we draw k cones with apex at u and angle $\frac{2\pi}{k}$. For each point $u \in S$ and cone C, we add an edge between u and the closest point to u in C. If one chooses k sufficiently large, the Yao graph becomes a t-spanner.

In 2014, Barba et al. [3] introduced the *continuous Yao graph* as a variation of Yao graph. The continuous Yao graph $cY(\theta)$ with vertex set S and angle θ defined as follows: For each $p, q \in S$, we add an edge from p to q in $cY(\theta)$ if there exists a *θ-cone*, a cone with aperture θ, with apex at p such that q is the closest point to p inside this θ-cone. In continuous Yao graph, for each $u \in S$, we rotate a θ-cone with apex u around u continuously, and connect u to the closest point inside the θ-cone during this rotation. They showed that $cY(\theta)$ has stretch factor at most $1/(1 - 2sin(\theta/4))$ for $0 < \theta < 2\pi/3$. Unlike Yao graphs that always have a linear number of edges for any constant k, in the worst case, continuous Yao graphs may have a quadratic number of edges. Since $cY(\theta) \subseteq cY(\gamma)$ for any $\theta \geq \gamma$, the continuous Yao graphs are useful in potential applications that require scalability. Moreover, when some rotations apply on the input point set the continuous Yao graphs are invariant [3].

One of the useful properties of a network is *fault tolerance* that is after one or more vertices or edges fail, the remaining graph is still a good network of alive vertices. In particular, a graph $G = (S, E)$ is called *k-vertex fault-tolerant t-spanner* [10] for S, denoted by (k,t)-VFTS for a given real number $t \geq 1$ and non-negative positive integer k, if for each set $S' \subseteq S$ with cardinality of at most k, the graph $G \backslash S'$ is a t-spanner for $S \backslash S'$. Also, G is called a *k-edge fault-tolerant t-spanner* [10] for S, denoted by (k,t)-EFTS, if for each set $E' \subseteq E$ with cardinality at most k and for each pair of points p and q in S, the graph $G \backslash E'$ contains a path P between p and q with $|P| \leq t|P_S|$ where P_S is the shortest path between p and q in the graph $K_S \backslash E'$ in which that K_S is Euclidean complete graph on S. Levcopoulos et al. [10] for the first time considered the problem of constructing fault-tolerant spanners in Euclidean spaces efficiently. They proposed three algorithms that construct k-vertex fault-tolerant spanners. Some other works on the fault tolerant spanners have been done [7,11].

In 2009, Abam et al. [1] introduced the concept of *region-fault tolerant spanner* for planar point sets. For a fault region F and a geometric graph G on a point set S, assume $G \ominus F$ is the remaining graph after removing the vertices of G that lie inside F and all edges that intersect F. For a set $\mathcal{F}$ of regions in the plane, an $\mathcal{F}$-fault tolerant t-spanner is a geometric graph G on S such that for any region $F \in \mathcal{F}$, the graph $G \ominus F$ is a t-spanner for $\mathcal{G}_c(S) \ominus F$, where $\mathcal{G}_c(S)$ is the complete geometric graph on S. They showed for any set of n points in the plane and any family $\mathcal{C}$ of convex regions, one can construct a $\mathcal{C}$-fault tolerant spanner of size $O(n \log n)$ in $O(n \log^2 n)$ time.

In 2013, Alamdari et al. [2] introduced the concept of *self-approaching* and *increasing-chord* graph drawings. The problem is that, we are given a graph and we need to check if the graph has an self-approaching or increasing-chord embedding in the Euclidean space.

A geometric graph is self-approaching, if for every pair s and t of vertices of the graph, there exists a self-approaching path from s to t, denoted by *st-path*. A path from s to t is a self-approaching path if for each q on the path, not only the vertices, but any place of the path, if a point p starts at s and moves toward q, it always get closer to q in its movement. Also, a graph G is called increasing-chord if, for each pair u and v of its vertices, there exists a path between u and v such that the path is self-approaching both from u to v and from v to u. Obviously, an increasing-chord graph is a self-approaching graph.

In the geometric context, the position of the vertices of the graph is fixed, so we just want to know whether a given geometric graph is self-approaching or increasing-chord. One of the interesting properties of these graph is the following. It is known that the stretch factor (or dilation) of any self-approaching graph is at most 5.3332 [9] and the stretch factor of any increasing-chord graph is at most 2.094 [15].

Our Results. In this paper, we prove the following results. For the rest of the paper S denotes a set of n points in the plane.

1. The continuous Yao graph $cY(\theta)$ on S is a $\mathcal{C}$-fault-tolerant geometric t-spanner of S, where $0 < \theta < \pi/3$ and $t \geq \frac{1}{1 - 2\sin(\theta/2)}$.
2. For any $\theta \leq \pi$ the graph $cY(\theta) \ominus h$ on S is connected, where h is an arbitrary half-plane in the plane.
3. There is a set of n points in the plane and a convex region C such that for every $\theta \geq \frac{\pi}{3}$, $cY(\theta) \ominus C$ is not connected.
4. There is a set P of n points in the plane such that for some angles θ, Yao graph on P with parameter θ is not a self-approaching graph. Instead, the corresponding continuous Yao graph on P is a self-approaching graph.
5. For every $\theta > 0$, $cY(\theta)$ is not necessarily a self-approaching graph.

2 $cY(\theta)$ is Fault-Tolerant

In this section, we show that the continuous Yao graph $cY(\theta)$ for $0 < \theta < \pi/3$ and $t \geq \frac{1}{1 - 2\sin(\theta/2)}$ is a $\mathcal{C}$-fault-tolerant geometric t-spanner where $\mathcal{C}$ is the family of convex regions in the plane. Furthermore, we show that for every $\theta \leq \pi$ and every half-plane h, $cY(\theta) \ominus h$ is connected. Moreover, we show that there is a set of n points such that for a some convex region C, $cY(\theta) \ominus C$ is not connected for every $\theta \geq \frac{\pi}{3}$. We need the following lemmas.

Lemma 1 [4]. *Let a, b and c be three points such that $|ac| \leq |ab|$ and $\angle bac \leq \alpha < \pi$. Then*

$$|bc| \leq |ab| - (1 - 2sin(\alpha/2))|ac|.$$

Lemma 2 [1]. *A geometric graph G on S is a C-fault tolerant t-spanner if and only if it is an $\mathcal{H}$-fault-tolerant t-spanner, where $\mathcal{H}$ is the family of all half-planes.*

Now, we prove the following theorem:

Theorem 1. *Let θ be a real number with $0 < \theta < \pi/3$ and let t be a real number with $t \geq \frac{1}{1-2\sin(\theta/2)}$. For any point set S, the continuous Yao graph $cY(\theta)$ is a C-fault-tolerant geometric t-spanner.*

Proof. By Lemma 2, it is sufficient to prove that $cY(\theta)$ is an $\mathcal{H}$-fault-tolerant geometric t-spanner.

Let h be an arbitrary half-plane in $\mathcal{H}$. We must show that for each pair of points $p, q \in S$ outside h, there is a t-path between p and q in $cY(\theta) \ominus h$. The proof is by induction on the rank of distance $|pq|$.

For the base step, suppose that the pair p and q is the closest pair in $cY(\theta) \ominus h$. Without loss of generality, assume the distance between p and h is less than of equal to the distance between q and h. Since p and q are outside h and $\theta < \pi/2$, there is a θ-cone C_p with apex at p completely outside h such that $q \in C_p$. Since pair p and q is the closest pair, by the construction of $cY(\theta)$, the edge (p,q) must be in $cY(\theta) \ominus h$. Note that if both p and q have same distance to h, then we can choose both p and q as apex, but we have to choose the point such that the corresponding θ-cone does not intersect h.

For the induction hypothesis step, suppose for each pair $u, v \in S$ outside h with $|uv| < |pq|$ there is a t-path between u and v connecting them in $cY(\theta) \ominus h$.

Now we consider the induction step. Since p and q are outside h, there is a θ-cone C_p with apex at p such that $q \in C_p$ (here we assumed that p is closer than q to h).

Suppose that r is the closest point to p inside the cone C_p (see Fig. 1). Since $\theta < \pi/3$, $1 - 2\sin(\theta/2) > 0$, and also since $|pr| \leq |pq|$, by Lemma 1 we have $|rq| < |pq|$. Therefore, by the induction hypothesis, there is a t-path Q between r and q in $cY(\theta) \ominus h$. Now consider the path $P := \{(p,r)\} \cup Q$. Clearly the path P connects p and q, and P is in $cY(\theta) \ominus h$. Now by Lemma 1, we have

$$
\begin{aligned}
|P| &= |pr| + |Q| \\
&\leq |pr| + t|rq| \\
&\leq |pr| + t\left(|pq| - (1 - 2\sin(\theta/2))|pr|\right) \\
&= t|pq| + (1 - t(1 - 2\sin(\theta/2)))\,|pr| \\
&\leq t|pq| \qquad \left(\text{since } t \geq \frac{1}{1 - 2\sin(\theta/2)}\right).
\end{aligned}
$$

Thus P is a t-path in $cY(\theta) \ominus h$ between p and q. This completes the proof. $\square$

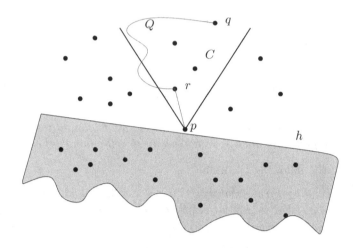

Fig. 1. Illustrating of the proof of Theorem 1.

Note that even though $cY(\theta)$ is a C-fault tolerant spanner, it is not necessarily connected. It is because in fault-tolerant spanners, we compare the graph with the complete graph. So if the complete graph after removing points in some regions becomes disconnected, then some of fault-tolerant spanners of the points set might be disconnected. In the following, we show that for every half-plane h in the plane, the graph $cY(\theta) \ominus h$ is a connected graph for every $\theta \leq \pi$.

Lemma 3. *The continuous Yao graph $cY(\pi)$ on each set S contains $CH(S)$, the convex hull of S.*

Proof. Proof is straightforward.   □

Theorem 2. *For any $\theta \leq \pi$ the graph $cY(\theta) \ominus h$ on S is connected, where h is a half-plane.*

Proof. It is easy to see that for every $\alpha, \beta > 0$, if $\alpha \geq \beta$ then $cY(\alpha) \subseteq cY(\beta)$. So, to prove the theorem, it is sufficient to show that $cY(\pi) \ominus h$ is connected for every half-plane h.

Let h be an arbitrary half-plane in the plane. Suppose that $cY(\pi) \ominus h$ is not connected.

Since $cY(\pi) \ominus h$ is disconnected, it has more than one connected component. At least one of the connected components of $cY(\pi) \ominus h$ contains a point from $CH(S)$. Suppose C and C' are two (distinct) connected components of $cY(\pi) \ominus h$ such that at least one of them, say C, contains a point from $CH(S)$. If both of C and C' contains a point from $CH(S)$ then by Lemma 3, part of $CH(S)$ which lies outside h is connected which is a contradiction because we assumed that C and C' are distinct connected components of $cY(\pi) \ominus h$.

Now, assume C' contains no vertices on $CH(S)$. Let x be the intersection of boundary of h and $CH(S)$ and let ℓ be a line through x and tangent to C' (see Fig. 2). We have two cases.

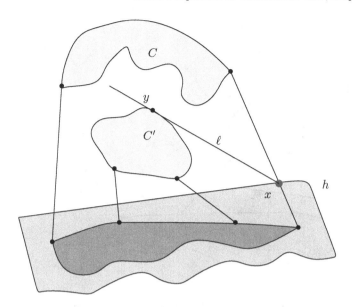

Fig. 2. Illustrating of the proof of Theorem 2: case 1.

Case 1: $\ell \cap C'$ contains exactly one point denoted by y.

If $x \notin S$ then clearly by the construction of graph $cY(\pi)$, the vertex y should be connected to a vertex of C that is contradiction, since we assumed that C and C' are distinct connected components of $cY(\pi) \ominus h$. Now, suppose that $x \in S$. Since $\ell \cap C'$ contains exactly one point, we can with continuously rotating the line ℓ around the point y (in Fig. 2, we rotate ℓ counter-clockwise), find a line ℓ' such that $\ell' \cap C'$ contains exactly one point that is y, and also the point in the intersection of ℓ' and boundary h does not belong to S. Hence, by the construction of graph $cY(\pi)$, point y should be connected to a vertex of C. That is a contradiction with that C and C' are distinct connected components of $cY(\pi) \ominus h$.

Case 2: ℓ is tangent to C' in at least two points.

We claim that there is a line ℓ' that is tangent to exactly one point of S on the convex hull of C' (denoted by s). Suppose that s be the farthest point of S on ℓ with respect to x, and suppose that r be the next point after s on the convex hull of C' in counterclockwise order (see Fig. 3). Let ℓ'' be a line that through of s and r, and let z be the intersection of ℓ'' and boundary of h. Now suppose that w be a point on the segment xz ($w \neq x, z$). Let ℓ' be a line that through of w and s. Clearly ℓ' is tangent to exactly one point of S that is s on the convex hull of C'. Now by the construction of $cY(\pi)$, vertex s should be connected to a vertex of connected component of C. This contradicts with the assumption that C and C' are distinct connected components of $cY(\pi) \ominus h$.

According to the above mentioned cases, $cY(\pi) \ominus h$ is connected. This completes the proof.   □

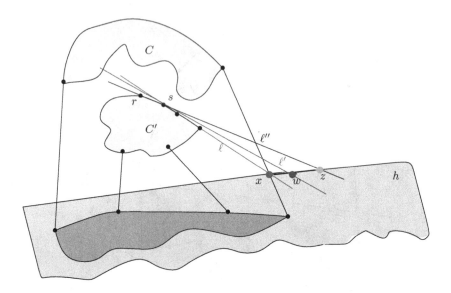

Fig. 3. Illustrating of the proof of Theorem 2: case 2.

In the following, we give an example such that for every $\theta \geq \frac{\pi}{3}$, $cY(\theta) \ominus C$ for some convex region C is not necessarily connected.

Assume $P_0 = \{(0,0),(1,\sqrt{3}),(2,0)\}$ which is the vertices of a equilateral triangle. Let P_i be the translation of P_0 by value $c \times i$ horizontally where c is a sufficiently large constant positive integer. Here, we consider $c = 10$ (see Fig. 4). We choose the set $P := \bigcup_{i=0}^{k-1} P_i$.

Let θ be an angle with $\theta \geq \frac{\pi}{3}$. Consider $cY(\theta)$ on P. Note that for $\theta \geq \frac{\pi}{3}$, $cY(\theta)$ on the vertices of an equilateral triangle is exactly the complete graph of their vertices. Now, let C be a convex region such that C only contains the points $(1,\sqrt{3})$, $(2,0)$, $(10,0)$ and $(11,\sqrt{3})$ from P (see Fig. 4). Since in the $cY(\theta)$ the vertex $(0,0)$ is only connected to the vertices $(1,\sqrt{3})$ and $(2,0)$, clearly in $cY(\theta) \ominus C$ there is no path between $(0,0)$ and $(12,0)$. Hence $cY(\theta) \ominus C$ is not connected.

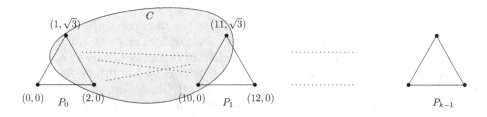

Fig. 4. An example that $cY(\theta)$ is disconnected after convex region fault.

3 $cY(\theta)$ is Not Self-approaching

In this section, we show that there is a set P of n points in the plane such that for some angles θ, Yao graph on P with parameter θ is not a self-approaching graph. Instead, the corresponding continuous Yao graph on P is a self-approaching graph. Finally, we show that there is a set of n points in the plane such that for every $\theta > 0$, continuous Yao graph $cY(\theta)$ is not a self-approaching graph.

In the following, we assume that the number of cones is even. With making some modifications, similar results holds for odd number of cones.

Now, let $p = (0,0), q = (1,0), r = (r_1, r_2)$ and $s = (s_1, s_2)$ be four points in the plane. For three points x, y and z, let $\angle yxz$ be the angle between the segment xy and xz. Now, assume that $\angle qpr = \alpha_1$, $\angle pqr = \alpha_2$, $\angle prq = \alpha_3$, $\angle qps = \beta_1$, $\angle pqs = \beta_2$, $\angle psq = \beta_3$, for some angles α_i and β_i with $0 < \alpha_i < \pi/2$ and $0 < \beta_i < \pi/2$ for $i = 1, 2, 3$ (see Fig. 5).

We call the quadrilateral on the four points p, q, r and s a *bad quadrilateral* if we have the following properties:

1. $|pq| \cos \alpha_1 < |pr| < |pq|$.
2. $|pq| \cos \alpha_2 < |qr| < |pq|$.
3. $|pq| \cos \beta_1 < |ps| < |pq|$.
4. $|pq| \cos \beta_2 < |qs| < |pq|$.

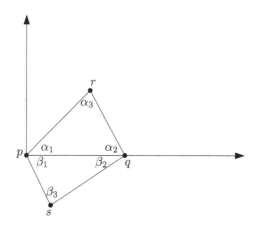

Fig. 5. A bad quadrilateral on the points p, q, r and s.

Now, we consider k cones, generated by the k-rays through origin, where the ith ray has angle $(i-1)\theta$ with the positive x-axis. The ith cones contains the points p such that the angle of line op with positive x-axis is bigger than or equal $(i-1)\theta$ and less than $i\theta$.

Let V be a set of four points $p = (0,0)$, $q = (1,0)$, $r = (r_1, r_2)$ and $s = (s_1, s_2)$ such that the quadrilateral on the vertices p, q, r and s is a bad quadrilateral.

It is easy to see that, for every θ with $\alpha_1 < \theta \leq \frac{\pi}{2}$ such that $\frac{2\pi}{\theta}$ is even number and $\beta_2 < \theta$, the Yao graph Y_k on V where $k = \frac{2\pi}{\theta}$ does not contain the edge $\{p, q\}$ but it contains the edges $\{p, r\}, \{p, s\}, \{q, r\}$ and $\{q, s\}$. So the Yao graph Y_k on the point set V is not a self-approaching graph, since every path from p to q is not a pq-path. Indeed, the following lemma shows that $cY(\theta)$ on V is a self-approaching graph.

Lemma 4. *Let $\alpha_1 + \beta_1 \geq \frac{\pi}{2}$ or $\alpha_2 + \beta_2 \geq \frac{\pi}{2}$. For every θ such that*

(a) $\alpha_1 < \theta \leq \frac{\pi}{2}$, and
(b) $\theta < \alpha_1 + \beta_1$ or $\theta < \alpha_2 + \beta_2$, and
(c) $\frac{2\pi}{\theta}$ is even number, and
(d) $\beta_2 < \theta$,

continuous Yao graph $cY(\theta)$ on V contains the edge $\{p, q\}$. Furthermore, $cY(\theta)$ on V is a self-approaching graph.

Proof. We prove the theorem for the case that $\theta < \alpha_1 + \beta_1$. Similar argument works for the case $\theta < \alpha_2 + \beta_2$.

Since $\theta < \alpha_1 + \beta_1$, by construction of $cY(\theta)$, clearly there is a θ-cone C with apex p that only contains the point q. Hence the edge $\{p, q\}$ is in $cY(\theta)$.

On the other hand, the Yao graph on V with angle parameter θ is a subgraph of $cY(\theta)$, so the edges $\{p, r\}, \{p, s\}, \{q, r\}$ and $\{q, s\}$ are in $cY(\theta)$. So, we only need to show that there is a self-approaching path from r to s and a self-approaching path from s to r.

Since $\alpha_1 + \beta_1 \geq \frac{\pi}{2}$ or $\alpha_2 + \beta_2 \geq \frac{\pi}{2}$, the path $r \to q \to s$ or $r \to p \to s$ is an rs-path and in the reverse direction is an sr-path. So there is an xy-path between all ordered pairs $\{x, y\}$ for $x, y \in V$. Hence, $cY(\theta)$ is a self-approaching graph on V. This completes the proof.   □

Next, we will give another point set such that the Yao graph on the point set is not self-approaching graph, but the continuous Yao graph is a self-approaching graph.

Let P be a set of n points as follows:

$$P = V \cup \{x_1, x_2, \ldots, x_{n-4}\}, \tag{1}$$

with $x_j = (c + j, 0)$ for $1 \leq j \leq n - 4$ where c is a sufficiently large constant positive integer. Here, we consider $c = 19$ (see Fig. 6).

Now, using Lemma 4 we can easily find the following result.

Lemma 5. *Let $\alpha_1 + \beta_1 \geq \frac{\pi}{2}$ or $\alpha_2 + \beta_2 \geq \frac{\pi}{2}$. For every θ such that*

(a) $\alpha_1 < \theta \leq \frac{\pi}{2}$, and
(b) $\theta < \alpha_1 + \beta_1$ or $\theta < \alpha_2 + \beta_2$, and
(c) $\frac{2\pi}{\theta}$ is even number, and
(d) $\beta_2 < \theta$,

continuous Yao graph $cY(\theta)$ on P is self-approaching and Y_k is not a self-approaching for $k = \frac{2\pi}{\theta}$.

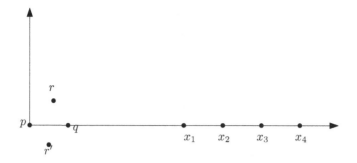

Fig. 6. The set P with $n = 8$.

Note that there are some point set that satisfy the conditions mentioned in above lemmas. For example, let $V = \{p, q, r, s\}$ with $p = (0,0), q = (1,0), r = (0.7839, 0.4422), s = (0.2161, -0.4422)$, and suppose that $k = 4$ and $\theta = \frac{\pi}{2}$. We can easily verify that the quadrilateral on the points of V is a bad quadrilateral. Moreover we have

$$\alpha_1 = 29.4271°, \alpha_2 = 63.9535°, \alpha_3 = 86.6198°,$$

$$\beta_1 = 29.4271°, \beta_2 = 63.9535°, \beta_3 = 86.6198°.$$

Now, suppose that $P = V \cup \{x_1, x_2, \ldots, x_{n-4}\}$ where $x_i = (19 + i, 0)$. This point set satisfies Lemma 5.

At the first view, it seems that the continuous Yao graph is a self-approaching graph, but this is not true in general. Next, we give a point set such that for each θ, the continuous Yao graph $cY(\theta)$ on the point set is not a self-approaching graph.

Theorem 3. *There is a set P of n points such that for every $\theta > 0$, $cY(\theta)$ on P is not a self-approaching graph.*

Proof. We prove the theorem for $0 < \theta \leq \frac{2\pi}{3}$. Since for every α_1 and α_2 with $\alpha_1 \leq \alpha_2$ we have $cY(\alpha_2) \subseteq cY(\alpha_1)$, the theorem holds for every $\theta > 0$.

Consider two points $p = (0,0)$ and $q = (1,0)$. Let C be a circle centered at the midpoint of the segment pq, with radius $\frac{1}{2}$. Let D_p and D_q be circles respectively centered at p and q with radius one. Let ℓ be the perpendicular bisector of segment pq (see Fig. 7).

Consider x and y as points outside C and inside $D_p \cap D_q$ such that $\angle xpq < \frac{\theta}{2}$ and $\angle ypq < \frac{\theta}{2}$. Let x' and y' be the symmetries of x and y with respect to line ℓ, respectively.

Now let $V = \{p, q, x, y, x', y'\}$, and consider $cY(\theta)$ on V. Since $\angle xpy = \angle x'qy' < \theta$, $cY(\theta)$ does not contain the edge $\{p, q\}$. So, by our selection of points x, y, x' and y', there is no self-approach path between p and q. Hence, $cY(\theta)$ on V is not self-approaching.

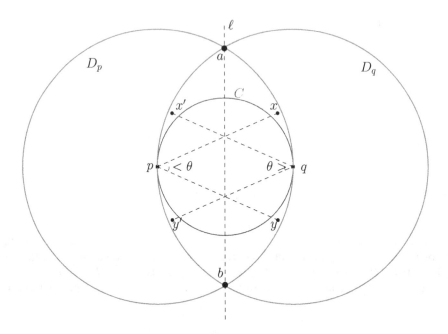

Fig. 7. Illustrating of the proof of Theorem 3.

One can extend this point set to a point set with arbitrary number of points such that $cY(\theta)$ is not a self-approaching graph of the point set. To this end, one can add any point to V which is sufficiently far from the points in the original set V, because adding points which are far can not help in making a self-approaching path from p to q. $\qquad\square$

4   Concluding Remarks

We proved that for $0 < \theta < \pi/3$ and $t \geq \frac{1}{1-2\sin(\theta/2)}$, the continuous Yao graph $cY(\theta)$ is a C-fault-tolerant geometric t-spanner. Furthermore, we showed that for every $\theta \leq \pi$ and every half-plane h, $cY(\theta) \ominus h$ is connected. Finally, we showed that for every $\theta > 0$, $cY(\theta)$ is not necessarily a self-approaching graph. The question whether $cY(\theta)$ for $\frac{\pi}{3} \leq \theta \leq \pi$ is a C-fault-tolerant geometric spanner with constant stretch factor remains open.

References

1. Abam, M.A., de Berg, M., Farshi, M., Gudmundsson, J.: Region-fault tolerant geometric spanners. Discrete Comput. Geom. **41**(4), 556–582 (2009)
2. Alamdari, S., Chan, T.M., Grant, E., Lubiw, A., Pathak, V.: Self-approaching graphs. In: Didimo, W., Patrignani, M. (eds.) GD 2012. LNCS, vol. 7704, pp. 260–271. Springer, Heidelberg (2013)

3. Barba, L., Bose, P., de Carufel, J.L., Damian, M., Fagerberg, R., van Renssen, A., Taslakian, P., Verdonschot, S.: Continuous Yao graphs. In: Proceedings of the 26th Canadian Conference on Computational Geometry, CCCG 2014, August 2014

4. Barba, L., Bose, P., Damian, M., Fagerberg, R., Keng, W.L., O'Rourke, J., van Renssen, A., Taslakian, P., Verdonschot, S., Xia, G.: New and improved spanning ratios for Yao graphs. In: Annual ACM Symposium on Computational Geometry, p. 30. ACM (2014)

5. Chandra, B., Das, G., Narasimhan, G., Soares, J.: New sparseness results on graph spanners. In: Proceedings of the Eighth Annual ACM Symposium on Computational Geometry, pp. 192–201. ACM (1992)

6. Chew, P.: There is a planar graph almost as good as the complete graph. In: Proceedings of the Second Annual ACM Symposium on Computational Geometry, pp. 169–177. ACM (1986)

7. Czumaj, A., Zhao, H.: Fault-tolerant geometric spanners. Discrete Comput. Geom. **32**(2), 207–230 (2004)

8. Eppstein, D.: Spanning trees and spanners. In: Handbook of Computational Geometry, pp. 425–461 (1999)

9. Icking, C., Klein, R., Langetepe, E.: Self-approaching curves. In: Mathematical Proceedings of the Cambridge Philosophical Society, vol. 125, pp. 441–453. Cambridge University Press, Cambridge (1999)

10. Levcopoulos, C., Narasimhan, G., Smid, M.: Improved algorithms for constructing fault-tolerant spanners. Algorithmica **32**(1), 144–156 (2002)

11. Lukovszki, T.: New results on fault tolerant geometric spanners. In: Dehne, F., Gupta, A., Sack, J.-R., Tamassia, R. (eds.) WADS 1999. LNCS, vol. 1663, pp. 193–204. Springer, Heidelberg (1999)

12. Lukovszki, T.: New results on geometric spanners and their applications. Ph.D. thesis, Heinz Nixdorf Institute and Department of Mathematics and Computer Science, Paderborn University, Paderborn, Germany (1999)

13. Narasimhan, G., Smid, M.: Geometric Spanner Networks. Cambridge University Press, Cambridge (2007)

14. Peleg, D., Schäffer, A.A.: Graph spanners. J. Graph Theor. **13**(1), 99–116 (1989)

15. Rote, G.: Curves with increasing chords. In: Mathematical Proceedings of the Cambridge Philosophical Society, vol. 115, pp. 1–12. Cambridge University Press, Cambridge (1994)

16. Smid, M.: Closest point problems in computational geometry. In: Handbook on Computational Geometry (1997)

17. Yao, A.C.C.: On constructing minimum spanning trees in k-dimensional spaces and related problems. SIAM J. Comput. **11**(4), 721–736 (1982)

Plane Geodesic Spanning Trees, Hamiltonian Cycles, and Perfect Matchings in a Simple Polygon

Ahmad Biniaz$^{(\boxtimes)}$, Prosenjit Bose, Anil Maheshwari, and Michiel Smid

Carleton University, Ottawa, Canada
ahmad.biniaz@gmail.com

Abstract. Let S be a finite set of points in the interior of a simple polygon P. A *geodesic graph*, $G_P(S, E)$, is a graph with vertex set S and edge set E such that each edge $(a, b) \in E$ is the shortest path between a and b inside P. G_P is said to be *plane* if the edges in E do not cross. If the points in S are colored, then G_P is said to be *properly colored* provided that, for each edge $(a, b) \in E$, a and b have different colors. In this paper we consider the problem of computing (properly colored) plane geodesic perfect matchings, Hamiltonian cycles, and spanning trees of maximum degree three.

1 Introduction

Let S be a set of n points in the interior of a simple polygon P with m vertices. For two points a and b in the interior of P, the *geodesic* $\pi(a, b)$, is defined to be the shortest path between a and b in the interior of P. A *geodesic graph*, $G_P(S, E)$, is a graph with vertex set S and edge set E such that each edge $(a, b) \in E$ is the geodesic $\pi(a, b)$ in P. If P is a convex polygon, then G_P is a straight-line geometric graph.

Let π_1 and π_2 be two, possibly self-intersecting, curves. We say that π_1 and π_2 *cross* if by traversing π_1 from one of its endpoints to the other endpoint we encounter a neighborhood of π_1 where π_2 intersects π_1 and switches from one side of π_1 to the other side [12]. We say that π_1 and π_2 are *non-crossing* if they do not cross. Two non-crossing curves can share an endpoint and can "touch" each other. If π_1 and π_2 are geodesics in a simple polygon, then they can intersect only once. They may have common line segments, but once they break apart, they do not meet again. See Fig. 1. A geodesic graph is said to be *plane* if the edges in E are pairwise non-crossing.

If the points in S are colored, then a geodesic graph G_P is said to be *properly colored* provided that, for each edge $(a, b) \in E$, a and b have different colors. For simplicity, in this paper we refer to a properly colored graph as a "colored graph". Let $\{S_1, \ldots, S_k\}$, where $k \geq 2$, be a partition of S. Let $K_P(S_1, \ldots, S_k)$ be the complete multipartite geodesic graph on S which has an edge between

Research supported by NSERC.

© IFIP International Federation for Information Processing 2016
Published by Springer International Publishing Switzerland 2016. All Rights Reserved.
M.T. Hajiaghayi and M.R. Mousavi (Eds.): TTCS 2015, LNCS 9541, pp. 56–71, 2016.
DOI: 10.1007/978-3-319-28678-5_5

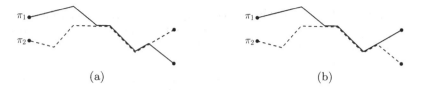

Fig. 1. (a) Two crossing geodesics, and (b) two non-crossing geodesics.

every point in S_i and every point in S_j, for all $1 \leq i < j \leq k$. Imagine the points in S to be colored, such that all the points in S_i have the same color, and for $i \neq j$, the points in S_i have a different color from the points in S_j. We say that S is a k-*colored* point set. Any colored geodesic graph, $G_P(S, E)$, is a subgraph of $K_P(S_1, \ldots, S_k)$.

If G_P is a perfect matching, a spanning tree, or a Hamiltonian cycle, we call it a *geodesic matching*, a *geodesic tree*, or a *geodesic Hamiltonian cycle*, respectively. A *colored geodesic matching* is a geodesic matching in $K_P(S_1, \ldots, S_k)$. Similarly, a *colored geodesic tree* (resp. a *colored geodesic Hamiltonian cycle*) is a geodesic tree (resp. geodesic Hamiltonian cycle) in $K_P(S_1, \ldots, S_k)$. A *plane colored geodesic matching* is a colored geodesic matching which is non-crossing. Similarly, a *plane colored geodesic tree* (resp. a *plane colored geodesic Hamiltonian cycle*) is a colored geodesic tree (resp. colored geodesic Hamiltonian cycle) which is non-crossing. Given a (colored) point set S in the interior of a simple polygon P, we consider the problem of computing a plane (colored) geodesic matching, geodesic Hamiltonian cycle, and geodesic 3-tree in $K_P(S_1, \ldots, S_k)$. A *t-tree* is a tree of maximum degree t. See Fig. 2.

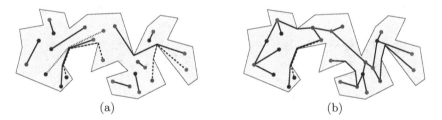

Fig. 2. (a) A plane colored geodesic matching, and (b) a plane colored geodesic 3-tree (Color figure online).

1.1   Preliminaries

We say that a set S of points in the pale is in *general position* if no three points of S are collinear. Moreover, we say that a set S of points in a simple polygon is *geodesically in general position* provided that, for any two points a and b in S, $\pi(a, b)$ does not contain any point of $S \setminus \{a, b\}$.

Toussaint [12] defined weakly-simple polygons—as a generalization of simple polygons—because in many situations concerned with geodesic paths the regions of interest are not simple but weakly-simple. A *weakly simple polygon* is defined as a closed polygonal chain $P = (p_1, \ldots, p_m)$, possibly with repeated vertices, such that every pair of distinct vertices of P partitions P into two non-crossing polygonal chains [12]. Alternatively, a closed polygonal chain P is weakly simple if its vertices can be perturbed by an arbitrarily small amount such that the resulting polygon is simple. From the computational complexity point of view, almost all data structures and algorithms developed for simple polygons work for weakly simple polygons with only minor modifications that do not affect the time or space complexity bounds. Hereafter, we consider a weakly simple polygon to be a simple polygon.

For two points a and b in the interior of a simple polygon P, $\pi(a, b)$ consists of a sequence of straight-line segments. We refer to a and b as the *external vertices* of $\pi(a, b)$, and refer to the other vertices of $\pi(a, b)$ as *internal vertices*. Moreover, we refer to the line segment(s) of $\pi(a, b)$ which are incident on a or b as the *external segments* and the other segments as *internal segments*. In the special case where $\pi(a, b)$ is a straight-line segment, $\pi(a, b)$ does not have any internal vertex nor any internal segment.

Observation 1. *The set of internal vertices of any geodesic in a simple polygon P is a subset of the reflex vertices of P.*

The *oriented geodesic*, $\overrightarrow{\pi}(a, b)$, is the geodesic $\pi(a, b)$ which is oriented from a to b. The *extended geodesic*, $\overline{\pi}(a, b)$, is obtained by extending the external segments of $\pi(a, b)$ till they meet the boundary of P. Let a' and b' be the points where $\pi(a, b)$ meet the boundary of P. Then, $\overline{\pi}(a, b)$ is equal to $\pi(a', b')$. An extended geodesic divides P into two (weakly) simple polygons. See Fig. 3.

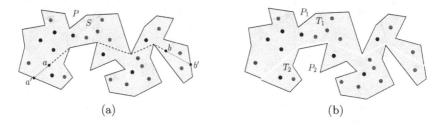

(a)   (b)

Fig. 3. A color-balanced point set S in the interior of a simple polygon P. (a) A balanced geodesic $\pi(a, b)$ with external vertices a and b. (b) The extended geodesic $\overline{\pi}(a, b)$ divides P into P_1, P_2, and partitions S into T_1, T_2 (Color figure online).

Assume S is partitioned into *color* classes, i.e., each point in S is colored by one of the given colors. S is said to be *color-balanced* if the number of points of each color is at most $\lfloor n/2 \rfloor$, where $n = |S|$. In other words, S is color-balanced if no color is in strict majority. Moreover, S is said to be *weakly color-balanced* if the number of points of each color is at most $\lceil n/2 \rceil$. Assume S is color-balanced and is in the interior of a simple polygon P. Let π be a geodesic in P. Let P_1 and P_2 be the (weakly) simple polygons on each side of the extended geodesic $\bar{\pi}$. Let T_1 and T_2 be the points of S in P_1 and P_2, respectively. We say that π is a *balanced geodesic* if both T_1 and T_2 are color-balanced and the number of points in each of T_1 and T_2 is at most $\frac{2n}{3} + 1$. See Fig. 3. The ham-sandwich geodesic (see [5]) is a balanced geodesic: given a set R of red points and a set B of blue points in a simple polygon P, a ham-sandwich geodesic is a geodesic which has its endpoints on the boundary of P and has at most $|R|/2$ red points and at most $|B|/2$ blue points on each side.

By Observation 1, both endpoints of any internal segment of a ham-sandwich geodesic are reflex vertices of P. Thus, we have the following observation:

Observation 2. *Let R and B be two disjoint sets of points in a simple polygon P. Let F be the set of reflex vertices of P. Let π be a ham-sandwich geodesic for R and B in P. If $R \cup B \cup F$ is in general position, then*

- *the internal segments of π do not contain any point of $R \cup B$.*
- *if $|R|$ (resp. $|B|$) is an even number, then the external segments of π do not contain any point of R (resp. B).*
- *if $|R|$ (resp. $|B|$) is an odd number, then exactly one of the external segments of π contains exactly one point of R (resp. B). Moreover, if both $|R|$ and $|B|$ are odd numbers, then the two points which are on π, belong to different external segments of π (assuming π is not a straight-line segment).*

Bose et al. [5] presented an $O((n+m) \log m)$ expected-time randomized algorithm for finding a ham-sandwich geodesic. Their algorithm is optimal in the algebraic computation tree model.

1.2 Non-crossing Structures in the Plane

Let S be a set of points in general position in the plane. Let $K(S)$ be the complete straight-line geometric graph on S. One can compute a plane Hamiltonian cycle in $K(S)$ in the following way. Let c be a point in $\mathbb{R}^2 \setminus S$ which is in the interior of the convex hull of S. Sort the points in S radially around c, then connect each point to its successor. The resulting structure, say H, is a plane Hamiltonian cycle in $K(S)$. By removing any edge from H a plane 2-tree is obtained. By picking every second edge of H a plane perfect matching is obtained (assuming $|S|$ is an even number).

Hereafter, assume S is partitioned into $\{S_1, \ldots, S_k\}$, where $k \geq 2$, and the points in S_i are colored C_i. Let $K(S_1, \ldots, S_k)$ be the complete straight-line multipartite geometric graph on S. Observe that if $K(S_1, \ldots, S_k)$ contains a plane Hamiltonian path, then S is weakly color-balanced. The reverse may not be true; if S is (weakly) color-balanced, it is not always possible to find a plane Hamiltonian path (or a plane 2-tree) in $K(S_1, \ldots, S_k)$. See [1,8] for examples. Kaneko [7] showed that if $k = 2$ and S is color-balanced, i.e., $|S_1| = |S_2|$, then $K(S_1, S_2)$ contains a plane 3-tree. Kano et al. [9] extended this result for $k \geq 2$: if S is weakly color-balanced, then $K(S_1, \ldots, S_k)$ contains a plane 3-tree.

A necessary and sufficient condition for the existence of a perfect matching (or colored matching) in $K(S_1, \ldots, S_k)$ follows from a result of Sitton [11].

Corollary 1. *Let* $\{S_1, \ldots, S_k\}$ *be a partition of a point set S in the plane, where* $k \geq 2$ *and* $|S|$ *is even. Then,* $K(S_1, \ldots, S_k)$ *has a colored matching if and only if S is color-balanced.*

Aichholzer et al. [2], and Kano et al. [9] show that the same condition as in Corollary 1 is necessary and sufficient for the existence of a plane colored matching in $K(S_1, \ldots, S_k)$:

Theorem 1 (Aichholzer et al. [2], and Kano et al. [9]). *Let* $\{S_1, \ldots, S_k\}$ *be a partition of a point set S in the plane, where* $k \geq 2$ *and* $|S|$ *is even. Then,* $K(S_1, \ldots, S_k)$ *has a plane colored matching if and only if S is color-balanced.*

In fact, they show something stronger. Aichholzer et al. [2] showed that any minimum-weight colored matching in $K(S_1, \ldots, S_k)$, which minimizes the total Euclidean length of the edges, is plane. Kano et al. [9] presented a constructive proof for the existence of a plane colored matching in $K(S_1, \ldots, S_k)$. Biniaz et al. [4] presented an algorithm which computes a plane colored matching in $K(S_1, \ldots, S_k)$ optimally in $\Theta(n \log n)$ time.

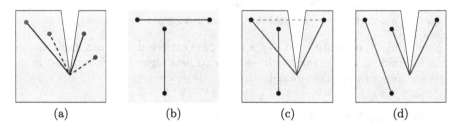

(a)   (b)   (c)   (d)

Fig. 4. (a) A minimum-weight colored geodesic matching which is crossing. (b) A non-crossing matching in the plane, and (c) its geodesic mapping which is crossing. (d) A non-crossing geodesic matching (Color figure online).

Although any minimum-weight colored matching in $K(S_1, \ldots, S_k)$ is non-crossing, this is not always the case for any minimum-weight colored geodesic matching in $K_P(S_1, \ldots, S_k)$, where the weight of a geodesic is defined to be the total Euclidean length of its line segments. Figure 4(a) shows a minimum-weight colored geodesic matching which is crossing.

As shown in Figs. 4(b)–(d) if we map a non-crossing matching in the plane to a geodesic matching inside a simple polygon, then the resulting matching may cross. This is also the case for non-crossing Hamiltonian cycles and non-crossing trees. Therefore, in order to compute a non-crossing geodesic structure in a simple polygon, it may not be an option to compute a non-crossing structure in the plane first, and then map it to a geodesic structure in the polygon.

1.3   Our Contributions

We generalize the notion of non-crossing (colored) structures for the case when the points are in the interior of a simple polygon and the edges are geodesics. Note that the problem of computing a non-crossing (colored) structure for points in the plane is the special case when the simple polygon is convex.

Let S be a set of n points in a simple polygon P with m vertices. Let $K_P(S)$ be the complete geodesic graph on S. In Sect. 2, we show that $K_P(S)$ contains a plane geodesic Hamiltonian cycle. This also proves the existence of a plane geodesic matching and a plane geodesic 2-tree in $K_P(S)$. We show how to construct such a cycle in $O(m + n \log(n + m))$ time.

Let $\{S_1, \ldots, S_k\}$, where $k \geq 2$, be a partition of S. Imagine the points in S to be colored, such that all the points in S_i have the same color, and for $i \neq j$, the points in S_i have a different color from the points in S_j. In Sect. 3 we extend the result of Kano et al. [9] for geodesic 3-trees. We show that if S is weakly color-balanced and $S \cup F$ is in general position, then $K_P(S_1, \ldots, S_k)$ contains a plane geodesic 3-tree and it can be computed in $O(nm + n^2 \log(n + m))$ time. In Sect. 4, we prove that if S is color-balanced and $S \cup F$ is in general position, then there exists a balanced geodesic for S in P. Moreover, if $|S|$ is even, then there exists a balanced geodesic which partitions S into two point sets each of even size. In either case, a balanced geodesic can be computed in $O((n + m) \log m)$ time. In Sect. 5 we compute a plane geodesic matching in $K_P(S_1, \ldots, S_k)$ in $O(nm \log m + n \log n \log m)$ time by recursively finding balanced geodesics.

2   Plane Geodesic Hamiltonian Cycles

2.1   Sweep-Path Algorithm

Let S be a set of n points in the plane. In a *sweep-line algorithm*, an imaginary vertical line scans the plane from left to right. The sweep line meets the points in S in the order determined by their x-coordinates. In a variant of the sweep-line algorithm, which is known as a *radial sweep algorithm*, an imaginary half-line, which is anchored at a point s in the plane, scans the plane in counter-clockwise order around s. The radial sweep meets the points in S in angular order around s. We extend the radial sweep algorithm for point set S in the interior of a simple polygon P. In the new algorithm, which we call *sweep-path algorithm*, an imaginary path which is anchored at a vertex s of P, scans P in "counter-clockwise" order around s. It gives a "radial ordering" for the points in S.

The sweep-path algorithm runs as follows. Let s be a vertex of P such that $S \cup \{s\}$ is geodesically in general position. Let t be a point which is initially at s. The algorithm moves t, in counter-clockwise order, along the boundary of P. See Fig. 5(a). At each moment the sweep-path is the oriented geodesic $\overrightarrow{\pi}(s, t)$. The algorithm stops as soon as t reaches its initial position, i.e., s. For two points $a, b \in S$ we say that $a \prec b$ if $\overrightarrow{\pi}(s, t)$ meets a before b. Thus, the sweep-path algorithm defines a total ordering $\mathcal{S} = (s_1, \ldots, s_n)$ on the points in S such that

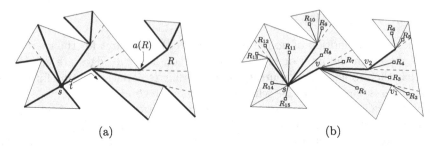

(a)   (b)

Fig. 5. (a) The shortest path tree rooted at s (in bold) which is extended (by dashed lines) to form the shortest path map for s. (b) The skeleton tree of SPT(s) (in bold) which is enhanced by the vertices representing the regions in SPM(s).

$s_i \prec s_j$, for all $1 \leq i < j \leq n$. See Fig. 7(a). We show how to obtain $\mathcal{S}$ in $O(m + n\log(n + m))$ time, where m is the number of vertices of P.

Let s be a vertex of P such that $S \cup \{s\}$ is geodesically in general position. We start by constructing the *shortest path tree* for s, denoted by SPT(s). This tree is defined to be the union of the shortest paths from s to all vertices of P. Then, we construct the *shortest path map* for s, denoted by SPM(s). The shortest path map for s is an enhancement of the shortest path tree rooted at s. See Fig. 5(a). Whereas the shortest path tree encodes the shortest path from every vertex of P to s, the shortest path map encodes the shortest path from every point inside P to s. Given SPT(s), the SPM(s) can be produced by partitioning the funnels of all edges of P in SPT(s). For each edge of P, we partition the funnel associated with it by extending the funnel edges. This partitions the funnel into triangular sectors (regions), each with a distinguished vertex called *apex*. The resulting subdivision is SPM(s). For a particular triangular region R in SPM(s) let $a(R)$ denote the apex of R (Fig. 5(a)). For any point p inside R the *predecessor* of p along $\overrightarrow{\pi}(s, p)$ is $a(R)$. Moreover, all points of R have the same internal vertex sequence in their shortest path to s.

Let T be the skeleton tree obtained from SPT(s) by removing its leaves (Fig. 5(b)). T contains the apex of all regions in SPM(s). For each region R in SPM(s) create a vertex which represents R, then, connect that vertex as a child to $a(R)$ in T. See Fig. 5(b). We order the children of each internal vertex $v \in T$ as follows. Let $P(v)$ be the union of the regions having v as their apex. See Fig. 6. Note that $P(v)$ is the union of a sequence of adjacent triangular regions all anchored at v, where v is a vertex of the boundary of $P(v)$. We order the children of v in counter-clockwise order.

We run depth-first-search on T to obtain an ordering $\mathcal{R} = (R_1, R_2, \dots)$ on the regions of SPM(s). See Figs. 5(b) and 6. Then, we locate the points of S in SPM(s). For each region R in SPM(s), let $L(R)$ be the list of points of S within R which are sorted counter-clockwise around $a(R)$. By replacing each R_i in $\mathcal{R}$ with $L(R_i)$ the desired ordering $\mathcal{S}$ is obtained. See Fig. 7(a).

SPM(s) has $O(m)$ size and can be computed in $O(m)$ time in a triangulated polygon using the algorithm of Guibas et al. [6]. A planar point location data

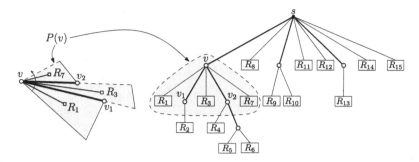

Fig. 6. The skeleton tree T (in bold) which is enhanced by the vertices representing the regions of SPM(s). For each $v \in T$, the children of v are ordered counter-clockwise.

structure for SPM(s) can be constructed in $O(m)$ time and answers point location queries in $O(\log m)$ time [10]. Thus, we can locate the points of S in SPM(s) in $O(m + n \log m)$ time. Making T to be an ordered tree takes $O(m)$ time by the construction of SPM(s) [6]. Sorting the points of S takes $O(n \log n)$ time for all regions. The depth first search algorithm runs in $O(m)$ time, and substituting each R_i with $L(R_i)$ takes $O(m + n)$ time. Thus, the total running time of the sweep-path algorithm is $O(m + n \log(n + m))$.

See the full version of the paper for the proof of the following lemma.

Lemma 1. *Let* $S = (s_1, \ldots, s_n)$ *be the ordering of the points in S obtained by the sweep-path algorithm. Let s_i, s_j, s_k and s_l be points in S such that $1 \leq i < j \leq k < l \leq n$. Then, $\pi(s_i, s_j)$ and $\pi(s_k, s_l)$ are non-crossing.*

2.2 Plane Geodesic Hamiltonian Cycles

Given a set S of n points in a simple polygon P with m vertices, in this section we show how to compute a plane geodesic Hamiltonian cycle on S.

A set $Q \subseteq P$ is called *geodesically* (or *relative*) *convex* if for any pair of points $a, b \in Q$ the geodesic between a and b in P, also lies in Q. The *geodesic hull* (or *relative convex hull*) of S in P, denoted by $GH(S)$, is defined to be the smallest geodesically convex set in P that contains S. Toussaint [12] showed that the geodesic hull of S in P is a weakly simple polygon, and can be computed in $O(m + n \log(n + m))$ time. Since for any two points a and b in S, $\pi(a, b)$ lies in $GH(S)$, without loss of generality, we assume that $P = GH(S)$. Let s_0 be a point of S on $GH(S)$. We run the sweep-path algorithm for $S \setminus \{s_0\}$ in $GH(S)$. It gives an ordering $S = (s_1, \ldots, s_{n-1})$ for the points in $S \setminus \{s_0\}$. We compute the following geodesic Hamiltonian cycle C (see Fig. 7),

$$C = \{(s_i, s_{i+1}) : 1 \leq i \leq n - 2\} \cup \{(s_0, s_1), (s_0, s_{n-1})\}.$$

Note that s_1 and s_{n-1} are the neighbors of s_0 on $GH(S)$. Therefore, (s_0, s_1) and (s_0, s_{n-1}) are non-crossing and do not cross (s_i, s_{i+1}) for all $1 \leq i \leq n - 2$. In addition, by Lemma 1 for $1 \leq i < j \leq k < l \leq n - 1$, (s_i, s_j) and (s_k, s_l)

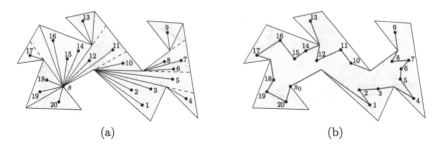

Fig. 7. (a) Points of S which are sorted by the sweep-path algorithm. (b) A plane geodesic Hamiltonian cycle (assuming s_0 is a point of S).

are non-crossing. This proves the planarity of C. By removing any edge from C, a plane geodesic 2-tree for S is obtained. By picking every second edge of C, a plane geodesic matching for S is obtained. Computing $GH(S)$ and running the sweep-path algorithm takes $O(m + n \log(n + m))$ time. Note that even if S is not geodesically in general position, one can compute C by simply modifying the sweep-path algorithm. Therefore, we have proved the following theorem:

Theorem 2. *Let S be a set of n points in a simple polygon with m vertices. Then, a plane geodesic Hamiltonian cycle, a plane geodesic 2-tree, and a plane geodesic matching for S can be computed in $O(m + n \log(n + m))$ time.*

3   Plane Geodesic Trees

Let S be a set of n points in the interior of a simple polygon P with m vertices. Let $\{S_1, \ldots, S_k\}$ be a partition of S, where the points in S_i are colored C_i. In this section we show that if S is weakly color-balanced and geodesically in general position, then there exists a plane colored geodesic 3-tree on S.

If $k \geq 4$, then by using the technique in the proof of Lemma 2 in [4], in $O(n)$ time we can reduce S to a weakly color-balanced point set with three colors such that any plane colored geodesic tree on the resulting 3-colored point set is also a plane colored geodesic tree on S. Therefore, from now on we assume that S is weakly color-balanced and its points colored by two or three colors. Let $CH(S)$ denote the convex hull of S. For a (geodesic) tree T and a given vertex s in T, let $d_T(s)$ denote the degree of s in T. Kano et al. [9] proved the following lemma and theorems for colored points in the plane. We adjusted the statements according to our setting and definitions.

Lemma 2 (Kano et al. [9]). *Let $(s_1, \ldots, s_n)$ be a sequence of $n \geq 3$ points colored with at most 3 colors[1] such that s_1 and s_n have the same color. If $\{s_1, \ldots, s_n\}$ is weakly color-balanced, then there exists an even number p, $2 \leq p \leq n - 1$, such that both $\{s_1, \ldots, s_p\}$ and $\{s_{p+1}, \ldots, s_n\}$ are weakly color-balanced.*

---

[1] Actually, they prove the statement of the theorem for 2- and 3-colored point sets.

Theorem 3 (Kano et al. [9]). *Let S be a set of points in general position in the plane which are colored red and blue. Let R be the set of red points and B the set of blue points. Let s be a vertex of $CH(S)$. If one of the following conditions holds, then there exists a plane colored 3-tree, T, on S such that $d_T(s) = 1$.*

(i) $|B| = 1$, $1 \le |R| \le 3$, and $s \in R$,
(ii) $2 \le |B|$, $|R| = |B| + 2$, and $s \in R$,
(iii) $2 \le |B| \le |R| \le |B| + 1$.

Theorem 4 (Kano et al. [9]). *Let S be a weakly color-balanced point set in general position in the plane which is colored by three colors. Let s be a vertex of $CH(S)$. Then, there exists a plane colored 3-tree, T, on S such that $d_T(s) = 1$.*

We extend Theorems 3 and 4 to prove the existence of plane geodesic trees on the colored points in the interior of a simple polygon. We adjust the proofs given in [9] to our setting, skipping the details.

Theorem 5. *Let S be a set of n points which is geodesically in general position in a simple polygon P with m vertices. Assume the points in S are colored red and blue. Let R be the set of red points and B the set of blue points. Let s be a vertex of $GH(S)$. If one of the following conditions holds, then in $O(nm + n^2 \log(n+m))$ time, one can compute a plane colored geodesic 3-tree, T, with vertex set S in P such that T is rooted at s and $d_T(s) = 1$.*

(i) $|B| = 1$, $1 \le |R| \le 3$, and $s \in R$,
(ii) $2 \le |B|$, $|R| = |B| + 2$, and $s \in R$,
(iii) $2 \le |B| \le |R| \le |B| + 1$.

Proof. The proof is by construction. Since for any two points a and b in S, $\pi(a, b)$ lies in $GH(S)$, without loss of generality, we may assume that $P = GH(S)$. If Condition (i) holds, the proof is trivial. Hence, assume that (ii) or (iii) holds. Let x and y be the left and the right neighbors of s on the boundary of $GH(S)$. If s and a neighboring vertex, say x, have distinct colors, then let T_1 be the tree obtained recursively on $S \setminus \{s\}$ which is rooted at x. Observe that x is a vertex of $GH(S \setminus \{s\})$ and $\pi(s, x)$ does not intersect $GH(S \setminus \{s\})$. Then, $T = T_1 + \pi(s, x)$ is the desired tree.

If s, x, and y have the same color, then let $\mathcal{S} = (s_1, \ldots, s_{n-1})$, where $s_1 = x$ and $s_{n-1} = y$, be the ordering of the points in $S \setminus \{s\}$ obtained by the sweep-path algorithm around s. See Fig. 8(a). If $s \in B$, then let $\mathcal{S} = (s_1, \ldots, s_n)$, where $s_1 = s$, $s_2 = x$, and $s_n = y$. See Fig. 8(b). In either case—$s \in R$ or $s \in B$—by Lemma 2 there exists an element s_p, with p even, such that if S_1 and S_2 be the points of S on each side of $\overline{\pi}(s, s_p)$ (not including s and s_p), then both $S_1 \cup \{s_p\}$ and S_2 are weakly color-balanced. Moreover, each of $S_1 \cup \{s_p\}$ and $S_2 \cup \{s_p\}$ satisfies one of the conditions (i), (ii), or (iii). Observe that $\pi(s, s_p)$ does not cross any of $GH(S_1 \cup \{s_p\})$ and $GH(S_2 \cup \{s_p\})$. In addition, s_p is a vertex of both $GH(S_1 \cup \{s_p\})$ and $GH(S_2 \cup \{s_p\})$. Let T_1 (resp. T_2) be the tree obtained recursively on $S_1 \cup \{s_p\}$ (resp. $S_2 \cup \{s_p\}$) which is rooted at s_p. Since $d_{T_1}(s_p) = 1$ and $d_{T_2}(s_p) = 1$, $T = T_1 + T_2 + \pi(s, s_p)$ is the desired tree.

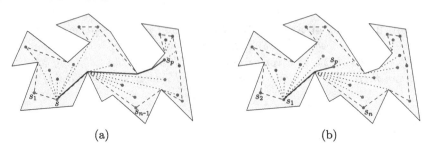

Fig. 8. Illustration of Theorem 5: (a) $|R| = |B| + 2, s \in R$, and (b) $|R| = |B| + 1, s \in B$ (Color figure online).

Computing the geodesic hull and running the sweep-path algorithm take $O(m + n \log(n + m))$ time. In the worst case, we recurse $O(|S|)$ times. Thus, the total running time of the algorithm is $O(nm + n^2 \log(n + m))$.   □

Theorem 6. *Let S be a 3-colored point set of size n which is geodesically in general position in a simple polygon P with m vertices. Let s be a vertex of $GH(S)$. If S is weakly color-balanced, then in $O(nm + n^2 \log(n + m))$ time, we can compute a plane colored geodesic 3-tree, T, with vertex set S in P such that T is rooted at s and $d_T(s) = 1$.*

Proof. Assume the points in S are colored red, green, and blue. Let R, G, and B be the set of red, green, and blue colors, respectively. Assume that $|B| \leq |G| \leq |R|$. The proof is by construction. If $|R| = \lceil |S|/2 \rceil$, we assume that G and B have the same color and solve the problem by Theorem 5. Assume that $|R| \leq \lceil |S|/2 \rceil - 1$. Observe that in this case $S \setminus \{s\}$ is weakly color-balanced. Let x and y be the left and the right neighbors of s on the boundary of $GH(S)$. If s and a neighbor vertex, say x, have distinct colors, then let T_1 be the tree obtained recursively on $S \setminus \{s\}$ which is rooted at x. Observe that x is a vertex of $GH(S \setminus \{s\})$ and $\pi(s, x)$ does not intersect $GH(S \setminus \{s\})$. Then, $T = T_1 + \pi(s, x)$ is the desired tree.

If s, x, and y have the same color, then let $\mathcal{S} = (s_1, \ldots, s_{n-1})$, where $s_1 = x$ and $s_{n-1} = y$, be the ordering of points in $S \setminus \{s\}$ obtained by the sweep-path algorithm on s. By Lemma 2 there exists an element s_p, with p even, such that if S_1 and S_2 be the points of S on each side of $\overline{\pi}(s, s_p)$, then both $S_1 \cup \{s_p\}$ and S_2 are weakly color-balanced. Since $|R| \leq \lceil |S|/2 \rceil - 1$, $S_2 \cup \{s_p\}$ is also color-balanced. Moreover, $\pi(s, s_p)$ does not cross any of $GH(S_1 \cup \{s_p\})$ and $GH(S_2 \cup \{s_p\})$. Let T_1 (resp. T_2) be the tree obtained recursively on $S_1 \cup \{s_p\}$ (resp. $S_2 \cup \{s_p\}$) which is rooted at s_p. Since $d_{T_1}(s_p) = 1$ and $d_{T_2}(s_p) = 1$, $T = T_1 + T_2 + \pi(s, s_p)$ is the desired tree.

As in the proof of Theorem 5, the running time is $O(nm + n^2 \log(n + m))$. □

4   Balanced Geodesics

Let S be set of $n \geq 3$ points in the interior of a simple polygon P with m vertices. Let F be the set of reflex vertices of P. Let $\{S_1, \ldots, S_k\}$ be a partition of S, where the points in S_i are colored C_i. Assume S is color-balanced. Recall that a balanced geodesic has its endpoints on the boundary of P and partitions S into two point sets T_1 and T_2, such that both T_1 and T_2 are color-balanced and $\max\{|T_1|, |T_2|\} \leq \frac{2n}{3} + 1$. We prove that if $S \cup F$ is in general position, then there exists a balanced geodesic for S in P. In fact, we show how to find such a balanced geodesic in $O((n + m) \log m)$ time by using a similar idea as in [4].

Theorem 7 (Balanced Geodesic Theorem). *Let S be a color-balanced point set of $n \geq 3$ points which is in the interior of a simple polygon P with m vertices. Let F be the set of reflex vertices of P. If $S \cup F$ is in general position, then in $O((n + m) \log m)$ time we can compute a geodesic π such that*

1. *π does not contain any point of S.*
2. *π partitions S into two point sets T_1 and T_2, where*
    (a) both T_1 and T_2 are color-balanced,
    (b) both T_1 and T_2 contains at most $\frac{2}{3}n + 1$ points,
    (c) if n is even, then both T_1 and T_2 have an even number of points.

Proof. Let $\{S_1, \ldots, S_k\}$ be the partition of S such that the points in S_i are colored C_i. We differentiate between three cases when $k = 2$, $k = 3$, and $k \geq 4$.

If $k = 2$, then $|S_1| = |S_2|$. Without loss of generality assume the points in S_1 are colored red and the points in S_2 are colored blue. Let π be a ham-sandwich geodesic of S in P. By Observation 2, if $|S_1|$ and $|S_2|$ are even numbers then π does not contain any point of S and hence it is a desired balanced geodesic. If $|S_1|$ and $|S_2|$ are odd, then one of the external segments of π contains a red point, say r, and the other external segment contains a blue point, say b. We adjust the external segments of π (by slightly moving its external vertices on P) such that both r and b lie on the same side of π. If π is a straight line segment, then we move π slightly such that both r and b lie on the same side of π. In either case, π is a desired balanced geodesic.

If $k \geq 4$, then by using the technique in the proof of Lemma 2 in [4], in $O(|S|)$ time we can reduce S to a color-balanced point set with three colors such that any balanced geodesic for the resulting 3-colored point set is also a balanced geodesic for S. Therefore, from now on we assume that S color-balanced and its points are colored by three colors, i.e., $k = 3$.

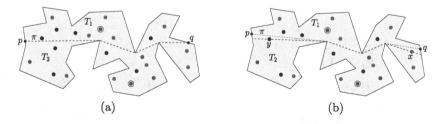

(a)   (b)

Fig. 9. Illustrating Theorem 7. The blue points in X are indicated by bounding circles. The ham-sandwich geodesic is in dashed lines. The geodesic π, with endpoints p and q, is a balanced geodesic when: (a) $|R|$ is even, and (b) $|R|$ is odd (Color figure online).

Let the points in S to be colored red, green, and blue. Let R, G, and B denote the set of red, green, and blue points, respectively. Without loss of generality assume that $1 \leq |B| \leq |G| \leq |R|$. Since P is color-balanced, $|R| \leq \lfloor \frac{n}{2} \rfloor$. Let X be an arbitrary subset of B such that $|X| = |R| - |G|$; note that $X = \emptyset$ when $|R| = |G|$, and $|X| = |B|$ when $|R| = \frac{n}{2}$ (when n is even). Let $Y = B - X$. Let π be a ham-sandwich geodesic for R and $G \cup X$ in P (by imagining that the points in $G \cup X$ have the same color). Let T_1 and T_2 denote the set of points of S on each side of π; see Fig. 9(a). Let R_1, G_1, and $B_1 (= X_1 \cup Y_1)$ be the set of red, green, and blue points in T_1 such that $X_1 = X \cap T_1$ and $Y_1 = Y \cap T_1$. Similarly, we define R_2, G_2, B_2, X_2, and Y_2 as subsets of T_2.

If $|R|$ is an even number, then π does not contain any point of $R \cup G \cup X$. If π contains any point $y \in Y$, then by Observation 2, y is on an external segment of π. We adjust that external segment (by slightly moving its external vertex on either side) such that it does not contain any point of S. If $|R|$ is an odd number, then π contains a point $x \in R$ and a point $y \in G \cup X$; see Fig. 9(b). By Observation 2, x and y are on different external segments of π (unless π is a straight line segment). In this case, without loss of generality, assume $|B_2| \geq |B_1|$. We adjust the external segments of π slightly such that x and y lie on the same side as T_2, i.e., $T_2 = T_2 \cup \{x, y\}$ (if π is a straight-line segment, then we move π slightly such that $T_2 = T_2 \cup \{x, y\}$); see Fig. 9(b). We prove that π satisfies the statement of the theorem. In either case we have $|R_1| = \lfloor |R|/2 \rfloor$, $|R_2| = \lceil |R|/2 \rceil$, $|G_1| + |X_1| = |R_1|$, and $|G_2| + |X_2| = |R_2|$. Therefore,

$$|T_1| \geq |R_1| + |G_1| + |X_1| = 2\lfloor |R|/2 \rfloor,$$
$$|T_2| \geq |R_2| + |G_2| + |X_2| = 2\lceil |R|/2 \rceil. \tag{1}$$

By the ham-sandwich geodesic we have $|G_1| \leq |R_1|$. This and Inequality (1) imply that $|G_1| \leq |R_1| = \lfloor |R|/2 \rfloor \leq |T_1|/2$. Similarly, we have $|G_2| \leq |R_2| = \lceil |R|/2 \rceil \leq |T_2|/2$. In order to prove that T_1 and T_2 are color-balanced, we have to show that $|B_1| \leq |T_1|/2$ and $|B_2| \leq |T_2|/2$. Let t_1 and t_2 be the total number of red and green points in T_1 and T_2, respectively; that is $t_1 = |R_1 \cup G_1|$ and $t_2 = |R_2 \cup G_2|$. Then,

$$|T_1| = t_1 + |B_1| \quad \text{and} \quad |T_2| = t_2 + |B_2|. \tag{2}$$

In addition,

$$
\begin{aligned}
t_1 &= |R_1| + |G_1| & t_2 &= |R_2| + |G_2| \\
&= |R_1| + (|R_1| - |X_1|) & &= |R_2| + (|R_2| - |X_2|) \\
&\geq 2|R_1| - |X| & &\geq 2|R_2| - |X| \\
&= 2\lfloor |R|/2 \rfloor - (|R| - |G|) & &= 2\lceil |R|/2 \rceil - (|R| - |G|) \\
&= \begin{cases} |G| & \text{if } R \text{ is even} \\ |G| - 1 & \text{if } R \text{ is odd,} \end{cases} & &= \begin{cases} |G| & \text{if } R \text{ is even} \\ |G| + 1 & \text{if } R \text{ is odd.} \end{cases}
\end{aligned}
\tag{3}
$$

Recall that $|B| \leq |G|$. Equation (2) and Inequality (3) imply that $|B_2| \leq |T_2|/2$. If $|R|$ is an even number, then Eq. (2) and Inequality (3) imply that

$|B_1| \leq |T_1|/2$. If $|R|$ is an odd number, then by assumption we have $|B_1| \leq |B_2|$; this implies that $|B_1| \leq |B| - 1$. Again by Eq. (2) and Inequality (3) we have $|B_1| \leq |T_1|/2$. Therefore, both T_1 and T_2 are color-balanced.

Now we prove the upper bound on the sizes of T_1 and T_2. By Inequality (1) both $|T_1|$ and $|T_1|$ are at least $2\lfloor |R|/2 \rfloor$. This implies that,

$$\max\{|T_1|, |T_2|\} \leq n - 2\lfloor \frac{|R|}{2} \rfloor \leq n - 2(\frac{|R| - 1}{2}) \leq n - |R| + 1.$$

Since R is the largest color class, $|R| \geq \lceil \frac{n}{3} \rceil$. Therefore, $\max\{|T_1|, |T_2|\} \leq n - \frac{n}{3} + 1 = \frac{2n}{3} + 1$.

The ham-sandwich geodesic π for R and $G \cup X$ in P can be computed in $O((n+m) \log m)$ time. Adjusting the external segments of π takes constant time. Thus, the total running time is $O((n + m) \log m)$.

See the full version of the paper for the proof of case (c), when n is even. □

5   Plane Colored Geodesic Matchings

Let S be a set of n points, with n an even number, which is in the interior of a simple polygon P with m vertices. Let F be the set of reflex vertices of P. Let $\{S_1, \ldots, S_k\}$, where $k \geq 2$, be a partition of S such that the points in S_i are colored C_i. Assume S is color-balanced. In this section we show that if $S \cup F$ is in general position, then $K_P(S_1, \ldots, S_k)$ contains a plane colored geodesic matching. In fact we show how to compute such a matching. If $k \geq 4$, by the technique of Lemma 2 in [4], in $O(n)$ time we can reduce S to a color-balanced point set with three colors such that any plane colored geodesic matching on the resulting 3-colored point set is also a plane colored geodesic matching on S. Thus, we assume that S color-balanced and its points are colored by at most three colors.

As in Theorem 6, we can adjust the technique used by Kano et al. [9]—for computing a non-crossing colored matching in the plane—to our setting. As a result we can compute a plane colored geodesic matching for S in P in $O(nm + n^2 \log(n + m))$ time.

Now we present an algorithm that computes a plane colored geodesic matching by recursively applying Balanced Geodesic Theorem as follows. By Theorem 7, we can find a balanced geodesic π that partitions P into simple polygons P_1 and P_2 containing point sets T_1 and T_2 such that both T_1 and T_2 are color-balanced with an even number of points, and $\max\{|T_1|, |T_2|\} \leq \frac{2n}{3} + 1$. Let M_1 (resp. M_2) be a plane colored geodesic matching for T_1 (resp. T_2) in P_1 (resp. P_2). Since P_1 and P_2 are separated by π, $M_1 \cup M_2$ is a plane colored geodesic matching for S. Therefore, in order to compute a plane colored geodesic matching for S in P, we compute a balanced geodesic for S in P, and then recursively compute plane colored geodesic matchings for T_1 in P_1 and for T_2 in P_2.

Let $T(n, m)$ denote the running time of the recursive algorithm on S and P, where $|S| = n$ and $|P| = m$. By Theorem 7, the balanced geodesic π can be computed in $O((n+m) \log m)$ time. The size of each of P_1 and P_2 is at most the size of P, and hence the recursions take $T(|T_1|, m)$ and $T(|T_2|, m)$ time. Thus, the running time of the algorithm can be expressed by the following recurrence:

$$T(n, m) = T(|T_1|, m) + T(|T_2|, m) + O((n + m) \log m).$$

Since $|T_1|, |T_2| \le \frac{2n}{3} + 1$ and $|T_1| + |T_2| = n$, this recurrence solves to

$$T(n, m) = O(nm \log m + n \log n \log m).$$

Theorem 8. *Let S be a color-balanced point set of size n, with n even, in a simple polygon P with m vertices, whose reflex vertex set is F. If $S \cup F$ is in general position, then a plane colored geodesic matching for S in P can be computed in $\min\{O(nm + n^2 \log(n + m)), O(nm \log m + n \log n \log m)\}$ time.*

Remark 1. By using the geodesic-preserving polygon simplification method of [3], the running time of any algorithm presented in this paper as $O(f(n, m))$ can be stated as $O(m + f(n, r))$, where r is the number of reflex vertices of P.

Remark 2. In Sect. 3, in each recursion step we run the sweep-path algorithm to sort the points around s_p. Having a semi-dynamic data structure for maintaining the geodesic hull which supports point deletions in $O(\text{polylog}(nm))$ worst case time, we can avoid the repetitive sorting. This would improve the running time for computing a plane colored geodesic 3-tree and a plane colored geodesic matching to $O((n + m)\text{polylog}(nm))$.

References

1. Abellanas, M., Garcia-Lopez, J., Hernández-Peñalver, G., Noy, M., Ramos, P.A.: Bipartite embeddings of trees in the plane. Discrete Appl. Math. **93**(2–3), 141–148 (1999)
2. Aichholzer, O., Cabello, S., Monroy, R.F., Flores-Peñaloza, D., Hackl, T., Huemer, C., Hurtado, F., Wood, D.R.: Edge-removal and non-crossing configurations in geometric graphs. Discrete Math. Theor. Comput. Sci. **12**(1), 75–86 (2010)
3. Aichholzer, O., Hackl, T., Korman, M., Pilz, A., Vogtenhuber, B.: Geodesic-preserving polygon simplification. In: Cai, L., Cheng, S.-W., Lam, T.-W. (eds.) Algorithms and Computation. LNCS, vol. 8283, pp. 11–21. Springer, Heidelberg (2013)
4. Biniaz, A., Maheshwari, A., Nandy, S.C., Smid, M.: An optimal algorithm for plane matchings in multipartite geometric graphs. In: Dehne, F., Sack, J.-R., Stege, U. (eds.) WADS 2015. LNCS, vol. 9214, pp. 66–78. Springer, Heidelberg (2015)
5. Bose, P., Demaine, E.D., Hurtado, F., Iacono, J., Langerman, S., Morin, P.: Geodesic ham-sandwich cuts. Discrete Comput. Geom. **37**(3), 325–339 (2007)
6. Guibas, L.J., Hershberger, J., Leven, D., Sharir, M., Tarjan, R.E.: Linear-time algorithms for visibility and shortest path problems inside triangulated simple polygons. Algorithmica **2**, 209–233 (1987)
7. Kaneko, A.: On the maximum degree of bipartite embeddings of trees in the plane. In: Akiyama, J., Kano, M., Urabe, M. (eds.) JCDCG 1998. LNCS, vol. 1763, pp. 166–171. Springer, Heidelberg (2000)

8. Kaneko, A., Kano, M.: Discrete geometry on red and blue points in the plane—a survey. In: Aronov, B., Basu, S., Pach, J., Sharir, M. (eds.) Discrete and Computational Geometry. Algorithms and Combinatorics, pp. 551–570. Springer, Heidelberg (2003)

9. Kano, M., Suzuki, K., Uno, M.: Properly colored geometric matchings and 3-trees without crossings on multicolored points in the plane. In: Akiyama, J., Ito, H., Sakai, T. (eds.) JCDCGG 2013. LNCS, vol. 8845, pp. 96–111. Springer, Heidelberg (2014)

10. Kirkpatrick, D.G.: Optimal search in planar subdivisions. SIAM J. Comput. **12**(1), 28–35 (1983)

11. Sitton, D.: Maximum matchings in complete multipartite graphs. Furman Univ. Electron. J. Undergraduate Math. **2**, 6–16 (1996)

12. Toussaint, G.T.: Computing geodesic properties inside a simple polygon. Revue D'Intelligence Artificielle **3**(2), 9–42 (1989)

Visibility Graphs of Anchor Polygons

Hossein Boomari$^{(\boxtimes)}$ and Alireza Zarei

Department of Computer Science, Sharif University of Technology, Tehran, Iran
mh.hima@gmail.com

Abstract. Visibility graph of a polygon corresponds to its internal diagonals and boundary edges. For each vertex on the boundary of the polygon, we have a vertex in this graph and if two vertices of the polygon see each other there is an edge between their corresponding vertices in the graph. Two vertices of a polygon see each other if and only if their connecting line segment completely lies inside the polygon. Recognizing visibility graphs is the problem of deciding whether there is a simple polygon whose visibility graph is isomorphic to a given graph. Another important problem is to reconstruct such a polygon if there is any. These are well-known and well-studied, but yet open problems in geometric graphs and computational geometry. However, these problems have been solved efficiently for special cases where the target polygon is known to be a tower or a spiral polygon. In this paper, we solve these recognizing and reconstruction problems for another type of polygons, named *anchor polygons*.

Keywords: Visibility graph · Polygon reconstruction · Recognizing visibility graph · Anchor polygon

1 Introduction

Visibility graph of a simple planar polygon is a graph in which there is a vertex for each vertex of the polygon and for each pair of visible vertices of the polygon there is an edge between their corresponding vertices in this graph. Two points in a simple polygon are visible from each other if and only if their connecting segment completely lies inside the polygon. In this definition, each pair of adjacent vertices on the boundary of the polygon are assumed to be visible from each other. This implies that we have always a Hamiltonian cycle in a visibility graph which determines the order of vertices on the boundary of the corresponding polygon.

Computing the visibility graph of a given simple polygon has many applications in computer graphics [10], computational geometry [9] and robotics [11]. There are several efficient polynomial time algorithms for this problem [9].

This concept has been studied in reverse as well: Is there any simple polygon whose visibility graph is isomorphic to a given graph and if there is such a polygon, is there any way to reconstruct it (find positions for its vertices on

M.T. Hajiaghayi and M.R. Mousavi (Eds.): TTCS 2015, LNCS 9541, pp. 72–89, 2016.
DOI: 10.1007/978-3-319-28678-5_6

the plain)? The former problem is known as recognizing visibility graphs and the latter one is known as reconstructing polygon from visibility graph. Both these problems are widely open. The only known result about the computational complexity of these problems is that they belong to *PSPACE* [1] complexity class and more specifically belong to the class of *Existence theory of reals* [8]. This means that it is not even known whether these problems are *NP-Complete* or can be solved in polynomial time. Even, if we are given the Hamiltonian cycle of the visibility graph which determines the order of vertices on the boundary of the target polygon, the exact complexity class of these polygons are still unknown.

However, these problems have been solved efficiently for special cases of tower and spiral polygons. In these special cases, we know that the given graph and the Hamiltonian cycle correspond to a tower polygon or a spiral one. A tower polygon consists of two concave chains on its boundary whose share one vertex and the other end point of these chains are connected by a segment (See Fig. 1.a). A spiral polygon has exactly one concave and one convex chain on its boundary (See Fig. 1.b). The recognizing and reconstruction problems have been solved for tower polygons in linear time in terms of the size of the graph [3]. It has been shown in [3] that a given graph is the visibility graph of a tower polygon if and only if by removing the edges of the Hamiltonian cycle from the graph, an isolated vertex and a connected bipartite graph are obtained and the bipartite graph has *strong ordering* following the order of vertices in the Hamiltonian cycle. A strong ordering on a bipartite graph $G(V, E)$ with partitions U and W is a pair of $<_V$ and $<_W$ orderings on respectively U and W such that if $u <_U u'$, $w <_W w'$, and there are edges (u, w') and (u', w) in E, the edges (u', w') and (u, w) also exist in E. Graphs having such ordering are also called *strong permutation graphs*. The recognizing and reconstruction problems have been solved efficiently for spiral polygons, too [2]. Because we need this method in our algorithm, we describe this method in more details in Sect. 2.

Although there is a bit progress on the recognizing and reconstruction problems, there have been plenty of studies on characterizing visibility graphs [2,3,5,6,12–14]. In 1988, Ghosh introduced three necessary conditions for visibility graphs and conjectured their sufficiency [5]. In 1990, Everett proposed a graph that rejects Ghosh's conjecture [1]. He also refined the third necessary condition of Ghosh to a new stronger condition [6]. In 1992, Abello *et al.* built a graph satisfying Ghosh's conditions and the stronger version of the third condition which was not the visibility graph of any simple polygon [15] rejecting the sufficiency of these conditions. Again, in 1997, Ghosh added his forth necessary condition and conjectured that this condition along with his first two conditions and the stronger version of the third condition are sufficient for a graph to be a visibility graph. Finally, in 2005 Streinu proposed a counter example for this conjecture [7]. Alongside these tries in 1994, Abello *et al.* proposed four necessary constraints for a graph to be visibility graph of a polygon and they conjectured that these constraints are verifiable efficiently [17]. Later in 1995, Abello *et al.* showed these constraints are sufficient for recognizing and reconstruction of 2-spiral polygons[1], without much contribution to their computational complexity [16].

In this paper, we consider these problems for another type of polygons called *anchor polygons*. The boundary of an anchor polygon is composed of two concave chains and a convex one (See Fig. 1.c). We characterized these polygons with some efficiently realizable constraints and show that both recognizing and reconstruction problems for a given pair of visibility graph and Hamiltonian cycle of an anchor polygon can be solved in $O(n^2)$ time where n is the number of vertices of the graph, or equivalently, the number of vertices of the anchor polygon.

In the remainder of this paper, we first introduce the algorithm of solving reconstruction and recognizing problems for spiral polygons in Sect. 2. Also, in this section we present some preliminaries and definitions used in next sections. In Sect. 3, we give an overview of our algorithm and extract key features from the graph to be used in our reconstruction algorithm. In Sect. 4 we present the algorithm and analyze its efficiency in Sect. 5.

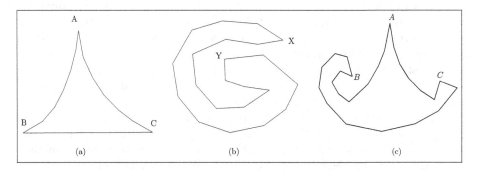

Fig. 1. (a) Tower polygon, (b) Spiral polygon, (c) anchor polygon.

2   Preliminaries and Definitions

In this section, we first briefly describe the recognizing and reconstruction algorithm for spiral polygons. We need these details in some parts of our algorithm. Then, we introduce some definition and basic facts to be used in next sections.

2.1   Spiral Polygons

Assuming that a pair of visibility graph and Hamiltonian cycle belong to a spiral polygon, Everret and Corneil proposed an efficient method to solve recognizing and reconstruction problems in these cases [2]. Here, we briefly describe their method.

---

[1] Polygons with at most 2 concave chains.

The visibility graph of a spiral polygon is a limited subclass of *interval graphs* [2]. This means that any interval graph satisfying another necessary condition corresponds to the visibility graph of a spiral polygon and vice versa (For the sake of brevity we skip describing this extra condition). This property helps to solve the recognizing problem for the visibility graphs of spiral polygons.

In any interval graph, and equivalently in the visibility graph of a spiral polygon, there are at least two vertices which form cliques with all their neighbors. Moreover, by removing one of these vertices, the remaining graph is still an interval graph. In a spiral polygon two joint vertices that connect its convex and concave chains have this property (form a clique with their neighbors) and by removing one of these vertices the residual graph will be another spiral polygon. Performing this *elimination scheme* from one of the joint vertices toward the other one will finally give us an ordered sequence of removed vertices and a subset of vertices which make a clique composed of the other joint vertex and its neighbors.

Assume that $< v_1, v_2, ..., v_k >$ is the ordered sequence of removing vertices and $\{v_{k+1}, ..., v_n\}$ is the set of remaining vertices which make a clique. Assume that v_n is a joint vertex and v_{k+1} is the only concave vertex in this set. We put $\{v_{k+1}, ..., v_n\}$ on the boundary of an arbitrary convex polygon. Then, the position of the vertices $< v_1, v_2, ..., v_k >$ are located in reverse order inductively as follows: Assume that v_c is the last located convex vertex before v_l and v_r is the last located concave vertex before v_l. For the induction base step, we set $v_r = v_n$ and $v_c = v_{k+2}$. To locate the position of v_{l-1} in an inductive step, assume that v_t is the closest convex vertex to v_r which sees v_{l-1}. By closest we mean that v_t is the first vertex on the Hamiltonian cycle when we move from v_r along the reconstructed part of the concave chain and then go along the reconstructed part of the convex chain. If v_l is a convex vertex, v_{l-1} is located somewhere in the angle $\widehat{v_t' v_r v_{t-1}'}$ (see Fig. 2.a) where $v_r v_t'$ (resp. $v_r v_{t-1}'$) is the half line from v_l along $v_r v_t$ (resp. $v_r v_{t-1}$) and in opposite side of v_t (resp. v_{t-1}) and in such a way that the new boundary do not cross itself. Otherwise (if v_l is a concave vertex), v_{l-1} is located somewhere in angle $\widehat{v_t' v_l v_{t-1}'}$ (see Fig. 2.b) where $v_l v_t$ and $v_l v_{t-1}$ half lines are defined similarly.

An important feature of this reconstruction algorithm is that starting from the initial convex polygon $v_{k+1}, ..., v_n$, the remainder of the spiral polygon can be reconstructed in an arbitrary close area of the concave vertex of this convex polygon. We use this feature in our reconstruction algorithm.

2.2 Definitions

In an anchor polygon, there are three specific vertices joining the three chains on the boundary of such a polygon. As shown in Fig. 1.c, the joint vertex between the concave chains is named A and the other joint vertices are named B and C. For simplicity, we assume that we have a left concave chain from A to B and a right concave chain from A to C and an underneath convex chain from B to C. We may refer to A as top joint vertex and to B and C as the left and right joint vertices, respectively. These names are consistent in all figures and inside the text to help readers have better perspective about the target anchor polygon.

We use term PQ^i for the ith vertex of the boundary of the polygon when we move from vertex P to vertex Q which both lies on the same chain. For example, AB^0 will be vertex A and AB^1 is the first vertex after A on the left concave chain. We also use $UV(P)$ as the closest vertex to U on chain UV which is visible to vertex P. In this notation, U and V can be any of the joint vertices A, B or C.

2.3   Basic Facts

From the convexity or concavity of the chains we have the following basic observations.

Observation 1. *Adjacent vertices (in the given Hamiltonian cycle) of each vertex $a \notin \{A, B, C\}$ of concave chains are not visible from each other.*

Observation 2. *Two vertices of the convex chain BC see each other if and only if no vertex of the concave chains blocks their visibility.*

Observation 3. *If a vertex v_i on the convex chain sees another vertex v_j on this chain, all vertices from v_i to v_j on this chain see each other. Moreover, if a vertex v_i on the convex chain does not see another vertex v_j on this chain and v_i is closer to B on chain BC, then none of the vertices from B to v_i see any one of the vertices from v_j to C.*

Observation 4. *The vertices AB^1 and AC^1 are always visible from each other.*

Observation 5. *At least one of the pairs of vertices (BC^1, BA^1) and (CB^1, CA^1) are visible from each other.*

Proof. The blocking vertices of (BC^1, BA^1) must lie on chain AC and blocking vertices of (CB^1, CA^1) must lie on chain AB. If BC^1 and BA^1 are not visible from each other, then none of the vertices of the chain AB can block the visibility of CB^1 and CA^1.   □

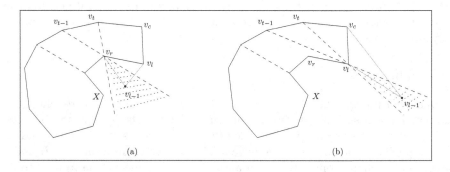

Fig. 2. The reconstruction algorithm for spiral polygons

3   Recognizing Algorithm: Determining Joint Vertices

We propose a constructive algorithm to solve both recognizing and reconstruction problems for anchor polygons. In this algorithm, we first determine the three chains on the given Hamiltonian cycle. During this process, some necessary conditions of the recognizing algorithm are verified. Then, the area of the target polygon is decomposed into four sub-polygons: a tower polygon, a convex polygon and two spiral polygons. The tower polygon is reconstructed first. Then, the convex polygon is constructed under the base edge of the built tower, and, finally the spiral polygons are built and attached to the sides of the constructed tower and convex polygons (See Fig. 5). Again, during this reconstruction process, the necessary conditions of the given visibility graph are checked to have a recognizing algorithm as well as the reconstruction one.

The details of the decomposition and reconstruction phases are described in Sect. 4. Here, we give a method for identifying the joint vertices A, B and C from which the three chains on the boundary of the target polygon are obtained. To do this, we first assume that we know vertex A and propose an algorithm for finding vertices B and C. Then, we propose a method for identifying candidate vertices for A.

3.1   Finding Joint Vertices B and C

When we move from A on the Hamiltonian cycle in both directions, from Observations 1 and 5 we can find at least one of the joint vertices B or C. This is the first visited vertex in these walks whose adjacent vertices in the Hamiltonian cycle see each other. Our algorithm for finding the other vertex is exactly the same: Walk along the Hamiltonian cycle from A in both directions until a vertex with this property (its adjacent vertices in Hamiltonian cycle see each other) is found in each direction. This algorithm will successfully find correct vertices as B and C if both pairs (BA^1, BC^1) and (CA^1, CB^1) are visible from each other. But, in some cases one of these pairs are not visible. Then, it seems that, our algorithm is failed to find joint vertex B or C.

We assume that both concave chains has at least one vertex other than the joint vertices. Otherwise, the target polygon will be a spiral one and can be recognized and reconstructed by algorithm proposed in [2]. Assume that $G(V, E)$ and H are the given pair of visibility graph and Hamiltonian cycle. The following theorem shows that the joint vertices B and C obtained by our algorithm along with A are the joint vertices of an anchor polygon whose visibility graph and Hamiltonian cycle are equivalent to the given pair of $G(V, E)$ and H if and only if $G(V, E)$ and H belongs to an anchor polygon.

Theorem 1. *Assume that for a given visibility graph $G(V, E)$ and Hamiltonian cycle H and a vertex A, the vertices B and C are the first visited vertices on H when we walk from A in both sides whose adjacent vertices see each other. Then, G and H correspond to an anchor polygon with top vertex A if and only if there is an anchor polygon with joint vertices A, B and C whose pair of visibility graph and Hamiltonian cycle are respectively isomorphic to G and H.*

Note that the proof, we proposed for this theorem is too long. Therefore, we present it in the Appendix section.

3.2   Determining Joint Vertex A

As a main part of our recognition algorithm, we describe a method for identifying the joint vertex A. If one of the concave chains AB or AC has only one edge (two vertices), the target polygon will be a spiral one and recognizing and reconstruction problems can be solved in such cases using the method proposed in [2]. Therefore, we assume that both chains AB and AC have at least one non-joint vertex. Then, the following observations are true for the joint vertex A.

Observation 6. *The pairs* (A, AB^2) *and* (A, AC^2) *do not see each other, but, the pair* (AB^1, AC^1) *are visible from each other.*

Observation 7. *All visible vertices from* A *see each other and along with* A *form a clique in the visibility graph.*

From these observations we have necessary conditions to find candidate vertices for A. We use these conditions in the first phase of our algorithm by moving along the Hamiltonian cycle and finding those vertices whose adjacent vertices see each other, but do not see vertices of distance 2 in the Hamiltonian cycle. For any one of the vertices passing this check we also check Observation 7. Then, we use our algorithm for finding other joint vertices (B and C) corresponding to any one of the candidate vertices for A. Clearly, for any candidate vertex p for A we must find corresponding joint vertices B_p and C_p where chain B_pC_p is convex. Each pair of visible vertices in convex chain B_pC_p must satisfy Observation 3. We show that there are at most three candidate vertices for A which pass the above conditions.

Assume that the given pair of visibility graph and Hamiltonian cycle belongs to an anchor polygon P with joint vertices A_P, B_P and C_P. Then, we have the following theorems.

Theorem 2. *If our algorithm find another candidate vertex for the top joint vertex* A *other than* A_P, B_P *and* C_P, *both chains* A_PB_P *and* A_PC_P *in* P *must lie completely on the same side of the line through vertices* $A_PB_P{}^1$ *and* $A_PC_P{}^1$.

Proof. Assume that a vertex v satisfies all conditions we check in our algorithm for finding candidate vertex A. While $v \notin \{A_P, B_P, C_P\}$, B_P and C_P are distinct vertices. This implies that v and its corresponding other joint vertices B_v and C_v (obtained by our algorithm for finding vertices B and C for a given vertex A) must lie on the convex chain B_PC_P of P. Then, both chains A_PB_P and A_PC_P must lie on the convex chain B_vC_v of the corresponding top joint vertex v. For the sake of a contradiction, assume that the chain A_PB_P or A_PC_P does not completely lie on one side of the line through $A_PB_P{}^1$ and $A_PC_P{}^1$. Without loss of generality (W.l.o.g.), assume that $A_PB_P{}^2$ lies below this line (See Fig. 3). From convexity of the chain B_PC_P, no vertex will block the visibility of $A_PB_P{}^2$

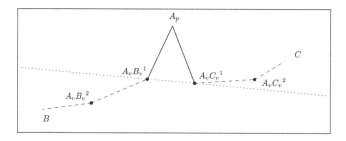

Fig. 3. Both chains $A_P B_P$ and $A_P C_P$ must completely lie on the same side of the line through $A_P B_P{}^1$ and $A_P C_P{}^1$.

and $A_P C_P{}^1$. On the other hand, these vertices lie on the convex chain $B_v C_v$ of the top joint vertex v. Therefore, according to Observation 3 A_P and $A_P B_P{}^2$ must also see each other which is a contradiction.   □

Theorem 3. *Our algorithm finds at most one candidate vertex for the top joint vertex A out of $\{A_P, B_P, C_P\}$. Moreover, if any one of the joint vertices B_P and C_P be a candidate vertex for A, there can be no more candidate vertex on the convex chain $B_P C_P$ out of B_P and C_P.*

Proof. For the sake of a contradiction, assume that our algorithm finds two candidate vertices A_1 and A_2 for the top joint vertex A out of $\{A_P, B_P, C_P\}$. As said in the proof of Theorem 2, both these vertices and their corresponding other joint vertices must lie on the convex chain $B_P C_P$ in P. W.l.o.g, assume that A_2 lies between B_P and A_1 on this convex chain (See Fig. 4). From the definition of joint vertices B and C and conditions for the top joint vertex, A_2 and $A_2 B_P{}^2$ must be invisible and $B_P A_P{}^1$ and $B_P C_P{}^1$ must be visible pairs. This forces that there must be at least one vertex between B_P and A_2 which means that $A_2 B_P{}^1$ can not be equal to B_P. While A_2 and $A_2 B_P{}^2$ are an invisible pair on the convex chain of P, there must be a blocking vertex b on $A_P B_P$ or $A_P C_P$ chains preventing their visibility. Clearly, b must be visible to A_2. On the other hand, both of the corresponding joint vertices of the top joint vertex A_1 (according to our algorithm) lies between vertices A_2 and C_P on the convex chain of P. This implies that all vertices of chain $B_P C_P$ from A_2 to B_P and vertices of the chains $A_P B_P$ and $A_P C_P$ in P lies on the convex chain of the candidate top joint vertex A_1. From Observation 3, when two vertices A_2 and b on this convex chain see each other, all vertices from A_2 to b, including $A_2 B_P{}^2$, must also see each other and form a clique which is a contradiction.

By the same argument we can prove that if the joint vertex C_P(or B_P) be a candidate for the top joint vertex, there cannot be any candidate vertex for the top joint vertex on the convex chain $B_P C_P$ out of $\{B_P, C_P\}$.   □

From the above theorems, we conclude that according to our algorithm there will be at most three candidates for the top joint vertex A. Precisely,

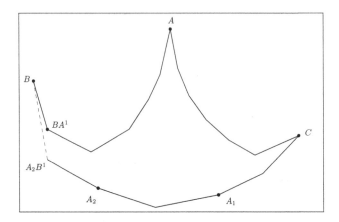

Fig. 4. A_2 and A_2B^2 must be invisible and A_2B^1 and A_2C^1 must be visible pairs.

if there was any other candidate other than A, either it is a vertex A' on B_PC_P ($A' \notin \{B_P, C_P\}$) or we have at most two candidates from B and C. From Observations 3 and 6 we can conclude that for the latter case the candidate vertices does not see any of the vertices of convex chain B_PC_P except the adjacent one in the Hamiltonian cycle.

4   Reconstruction Algorithm

In this section, we assume that we are given a pair of visibility graph, $G(V, E)$ and Hamiltonian cycle, H, and three joint vertices A, B and C and the goal is to obtain an anchor polygon $G(V, E)$ corresponding to these graph and cycle with A, B and C as its top, left and right joint vertices, respectively. Moreover, we assume that the visibility graph and the joint vertices satisfy conditions described in previous observations and conditions of previous algorithms (the joint vertices have been obtained by the algorithms described in Sect. 3). From previous section, we know that there are at most three options for these joint vertices. Therefore, to solve the recognizing algorithm we may run the following algorithm at most three times and if one of these runs leads to an anchor polygon it will be returned as a solution and if none of them produce a polygon it means that $G(V, E)$ and H do not belong to an anchor polygon.

Our reconstruction algorithm consists of two phases. Initially we decompose the target polygon into at most four regions and then these regions are reconstructed to build the final anchor polygon.

4.1   Anchor Polygon Decomposition

We define a line d as a bi-tangent line for both chains AB and AC if it passes through vertices M and M' on AB and AC, respectively, and both chains lie completely on the same side of it (See Fig. 5). From the visibility graph edges

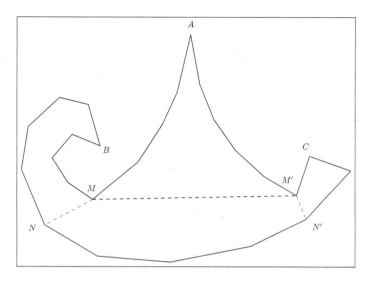

Fig. 5. Decomposition of an anchor polygon

we can find such a bi-tangent line: There is no edge from vertices of AM (resp. AM') to vertices of $M'C$ (resp. MB) except the edge MM'. Also, the polygon with boundary AM, AM' and edge MM' is a tower polygon.

Observation 8. *Each anchor polygon has exactly one bi-tangent line.*

Observation 9. *The bi-tangent line of an anchor polygon passes through its joint vertices B and C if and only if B and C see each other. In these cases, the convex chain BC lies completely on the opposite side of the bi-tangent line compare to A and all of the vertices of this convex chain are visible from each other, and so, they form a clique in the visibility graph.*

Let N and N' be vertices $BC(M')$ and $CB(M)$, respectively (See Fig. 5). As we stated before, the polygon with boundary vertices $< M, ..., B, ..., M' >$ is a tower polygon and polygon with boundary vertices $< M, N, ..., N', M' >$ is a convex one. Also, both polygons with boundary $< M, ..., B, ..., N >$ and $< M', ..., C, ..., N' >$ are spiral polygons and there is no edge between the vertices of one of them to the other one or the tower sub-polygon, except edges have an end point in $\{N, N', M, M'\}$ (See Fig. 5). Otherwise, we report that the pair $G(V, E)$ and H does not correspond to any anchor polygon with joint vertices A, B and C. Note that based of the shape of the anchor polygon, some of these four sub-polygons may not exist (it may be only a point or an edge).

This decomposition of the anchor polygon can be obtained from a given $G(V, E)$ and H and the three joint vertices A, B and C. After obtaining the bi-tangent line and the tower polygon as discussed above, the vertices N and N' are obtained from $G(V, E)$ and H according to their definition ($N = BC(M')$ and $N' = CB(M)$). Now (after checking the previous observations), because some of the vertices of these spiral sub-polygons may see some of the convex ones, we will extend boundary of them to $< M, ..., B, ..., N, ..., N', M' >$ and $< M', ..., C, ..., N', ..., N, M >$, respectively (clearly both of them are spiral polygons yet). The visibility graph of any one of these sub-polygons must satisfy the sub-polygon conditions. Precisely, the induced sub-graph of G on vertices of the tower polygon (resp. spiral polygons) must have necessary conditions of the visibility graph of a tower polygon (resp. spiral polygon) with these boundary vertices, and, the induced sub-graph of G on the convex sub-polygon must be a complete graph. Otherwise, we report that the pair $G(V, E)$ and H does not belong to an anchor polygon with the given joint vertices A, B and C.

4.2   Reconstructing Sub-polygons

Now, we are ready to propose the final step of our constructive algorithm for solving both recognizing and reconstruction problems. If we consider the union of the tower and convex sub-polygons, in the decomposition phase, as a single polygon, it will be an anchor polygon as well. But, this anchor polygon has this property that its bi-tangent passes through its non-top joint vertices B and C. We call such anchor polygons *simple anchor polygons*. The visibility graph of a simple anchor polygon with joint vertices A, B and C has the following properties.

Observation 10. *Each concave vertex of an anchor polygon sees one continues sub-chain of the convex chain.*

Observation 11. *The joint vertices B and C of a simple anchor polygon see the whole convex chain.*

Observation 12. *For each concave vertex p of a simple anchor polygon, the vertices of the convex chain which are visible to pA^1 are a subset of the vertices visible from p.*

Observation 13. *If both convex vertices p and pB^1 of a simple anchor polygon lie on the right side of the line through A and AB^1, the set of visible concave vertices from p is a subset of such set for pB^1 (See Fig. 6.a). Symmetrically, this is true for p and pC^1 if both lie on the left side of the line through A and AC^1.*

Observation 14. *Assume that $q = AB(p)$ is the closest vertex of the concave chain AB to A which is visible to a convex vertex p on the left side of the line through A to AC^1 in a simple anchor polygon. Then, none of the vertices of the sub-chain from A to $s = AC(q)A^1$ is visible from p and all vertices of the sub-chain from C to $t = AC(qB^1)$ are visible from p (See Fig. 6.b). It means that $AC(p)$ must be one of the vertices of the left concave chain from s to t.*

Symmetrically, for a convex vertex p lying on the right side of the line through A and AB^1 and $q = AC(p)$, $AC(p)$ must be one of the vertices of the right concave chain from $AB(q)A^1$ to $AB(qC^1)$.

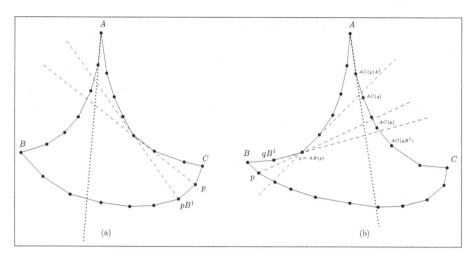

Fig. 6. (a) Visible points from p is a subset of the visible points from pB^1 (b) Visible and invisible vertices of concave chain AC from p

Trivially all above observations must be hold on the visibility graphs induced to the vertices of the tower and convex sub-polygons in our decomposition algorithm presented in Sect. 4.1. If these conditions hold, we reconstruct simple anchor polygon which corresponds to the obtained tower and convex sub-polygons.

To reconstruct a simple anchor polygon, we first reconstruct the tower polygon using the method presented in [3] (we use the method as a block-box procedure). Then the vertices of the convex chain are put on a convex curve from B to C supporting their order on the Hamiltonian cycle and the visibility graph constraints. To do this, we divide these vertices into these groups: The first group, called V_A, contains those vertices that see all vertices of both concave chains. From the above observations, these vertices must lie on the convex curve between the lines passing through A and AB^1, and A and AC^1 (See Fig. 7.a). The other groups are the sets V_B and V_C as shown in Fig. 7. To locate an arbitrary vertex $v \in V_B$ it must satisfy two conditions: assume that $p = AB(v)$ and $q = AC(v)$ are respectively, the top most vertices on chains AB and AC which are visible to v. According to the visibility graph constraints, p must lie on the left of the line through p and pA^1 and to the right of the line through p and q. Moreover, v does not see qA^1 and p is a blocking vertex for this invisibility. Therefore, if qA^1 is visible from p then v must lie to the left of the line through p and qA^1. Otherwise, as v lies to the left of the line through p and pA^1, the vertex p will block the visibility of v and qA^{-1} and there is no need to add more constraint to restrict position of v on the convex curve. Therefore, the intersection of the convex curve and this region must be non-empty. If this happens, we can put p on an arbitrary point of this part of the convex curve and for all points p that must be located in this region, we put them according to their order in

Hamiltonian cycle. As the last point of our algorithm, we must show that the intersection of the convex curve and constructed region of v is not empty. The region is restricted to lines d_1 and d_2 or lines d_2 and d_3 (See Fig. 7). It is simple to show that in both cases q is visible from p and in the latter one qA^1 is visible from p and lies above q. This implies that in both cases the region, and consequently, the intersection is not empty. Note that finding the corresponding regions and intersections for vertices of set V_C can be done similarly.

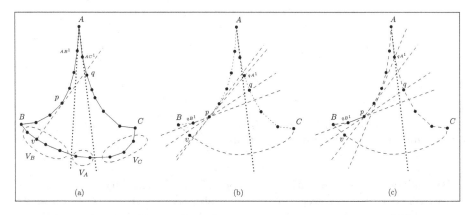

Fig. 7. (a) V_A, V_B and V_C (b) Invisibility of v and qA^1 needs to be cared (c) Invisibility of v and qA^1 do not need to be cared

After reconstructing the simple anchor polygon of the tower and the convex sub-polygons, we must build and attach the spiral sub-polygons to the sides of this simple anchor polygon. Recall that the boundary of the right and left sub-polygons are $< M, ..., B, ..., N, ..., N', M' >$ and $< M', ..., C, ..., N', ..., N, M >$, respectively. Moreover, remember that in these polygons there is no edge between the vertices of $< M, ..., B, ..., N >$ and $< M', ..., C, ..., N' >$ except edges that have an end point in $\{N, N', M, M'\}$. This helps to build these parts independently. Note that we can apply this independency by locating the remained vertices of these spiral polygons above the line through M and M'. We describe how to build the left spiral polygon and the right one can be built Symmetrically. If we apply the elimination scheme starting from the joint vertex B, we find a sequence of removed vertices, which includes all vertices of the left spiral polygon except vertices of the convex sub-polygon. Moreover, the remained vertices of this spiral polygon (which make a convex sub-polygon with respect to the Hamiltonian cycle) are the vertices of our convex sub-polygon $< M, M', N', N >$, which is already reconstructed. The spiral polygon reconstruction algorithm described in Sect. 2 says that we can start from an arbitrary convex polygon for the remained vertices and the sequence of the removed vertices can be put in an arbitrary small neighborhood of the only concave vertex (here it is M). This means that, by considering the convex sub-polygon as the starting convex polygon, we can reconstruct the left spiral polygon arbitrary close to M without intersecting the constructed tower polygon. Note that by using this method all vertices of the removed sequence are forced to be located above the line through M and M'.

5 Complexity Analysis

In this section we analyse the time complexity of our algorithm for recognizing and reconstruction of an anchor polygon from its visibility graph and Hamiltonian cycle.

Before beginning the analysis we assume that for each vertex we know its maximal cliques with its previous and successor vertices according to their order in Hamiltonian cycle separately. It means that for each vertex p we know how many vertices consecutively after (resp. before) p will make a clique with it as a number denoted by $C^+(p)$ (resp. $C^-(p)$). We can calculate these numbers in $O(n^2)$ for all the vertices using *Dynamic Programming* [4] where n is number of vertices in the visibility graph. In addition, we denote by $D^+(p)$ (resp. $D^-(p)$) the distance between p and the first visible vertex after (resp. before) its maximal clique, visible to p.

The first part of our algorithm finds candidate vertices. For this purpose our algorithm iterate on each vertex of the visibility graph for checking necessary conditions to be a candidate vertex. For each vertex p checking for visibility of the vertices adjacent to it in Hamiltonian cycle and checking for invisibility of (pB^1, pB^2) and (pC^1, pC^2) (B and C used here to illustrate vertices after and before vertex p) needs $O(1)$ for each vertex and $O(n)$ overally. Then, Finding vertex B_p and C_p will take $O(n)$ time for each vertex p and $O(n^2)$ overally. After finding these vertices, we should check for necessary sight condition between convex vertices which could be done in $O(n)$ for each vertex, using functions C^-, C^+, D^- and D^+, and $O(n^2)$ for all vertices. So, we can check necessary conditions for candidate vertices and finding them in $O(n^2)$ and will begin the reconstruction for each of them independently. As the number of candidate vertices are in $O(1)$, the time complexity of the reconstruction algorithm is the time required for one candidate set of A, B and C.

The reconstruction algorithm, take at most $O(n^2)$ for finding the bi-tangent line and then decomposing polygon into a simple anchor polygon and two spiral polygons. After reconstructing the tower polygon of that simple anchor polygon in $O(|E|)$ [3], it will take $O(n^2)$ for checking necessary and sufficient conditions for recognizing simple polygon and reconstructing it. Finally, recognizing each spiral polygon will take $O(n^2)$ using the algorithm of [2]. Summing all, our algorithm will recognize and reconstruct an anchor polygon in $O(n^2)$ time, where n is the number of vertices of the input visibility graph.

Appendix

Proof of Theorem 1

Proof. Trivially, the theorem is true when B and C are the joint vertices of the target polygon of G and H. Moreover, the theorem is trivially true when there is no anchor polygon with visibility graph G and Hamiltonian cycle H. Therefore, it is enough to prove the theorem for the cases where G and H belong to an anchor polygon P with joint vertex A and at least one of the other joint vertices of P is not in $\{B, C\}$. According to Observation 5, assume that C is a joint vertex of P and the other joint vertex is another vertex $B' \neq B$. This means that B' lies on the convex chain of P, and equal to a vertex BC^i where $i > 0$, while $B' = BC^0$ in anchor polygon P. We prove the theorem by induction on i. For $i = 1$, it means that when BA^1 and BC^1 do not see each other in P, we can consider the joint vertex B as a vertex of the left concave chain of P and considering BC^1 as the left joint vertex, and the visibility graph of this anchor polygon is still equivalent to G. This has been shown in Fig. 9 where (a) is the original polygon and (b) is the new one with BC^1 as a joint vertex.

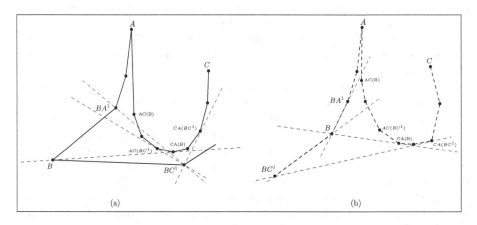

Fig. 8. Considering BC^1 as a joint vertex in (b) while BC^1 and BA^1 are invisible in (a).

To complete the proof we must show that it is always possible to locate BC^1 on the left concave chain of some anchor polygon without disturbing the visibility graph constraints. This is done by first proving that the induced visibility graph on vertices $U = < A, AB^1, ..., B, BC^1 >$ and $W = < A, AC^1, ..., CA(BC^1) >$ have strong ordering with partitions U and W (Then, we can build a tower polygon on these vertices in which B is a concave vertex on the left chain and BC^1 is the vertex on this concave chain) and second, proving that it is possible to place other vertices $< BC^2, ..., C >$ and $< CA(BC^1), ..., C >$ supplying the visibility graph edges.

For the first one, we know from P that the induced visibility graph on $U' = < A, AB^1, ..., B >$ and $W' = < A, AC^1, ..., CA(B) >$ have strong ordering. Then it is enough to consider the pairs (BC^1, w) and (u, w'), where $w, w' \in W$ and $w <^W w'$, w is visible from BC^1, and $u \in U$ is visible from w'. As shown in Fig. 8 if BC^1 sees w and w' is further than w from A, then u' must also be visible from BC^1, which means that (BC^1, w') exists in the graph. On the other hand, while

BC^1 and BA^1 are invisible from each other, u can only be the vertex B. Then, if $u = B$ sees w' it will also see all closer vertices to A than w' which are visible from BC^1. This complete the existence of the strong ordering on U and W.

For the second one, we show that after building the tower polygon on U and W, we can add remained vertices to find an anchor polygon with A, BC^1 and C as its joint vertices with the same visibility graph as P. These remaining vertices are the vertices of the right concave chain of P from C to $CA(BA^1)$ and the vertices of convex chain from C to BC^1. Later we will add the first set as a concave chain of a spiral polygon starts from $CABC^1$ and above the line passing from vertices BC^1 and $CA(BC^1)$. This is consistent with the visibility graph because none of these vertices see anyone of the vertices of the tower (See Fig. 8.b). The convex vertices from C to $BC^1 = B'$ are divided into three parts:

(V1) The visible vertices from B.
(V2) The visible vertices from B' but not visible to B.
(V3) The invisible vertices from B'.

It is simple to check that none of these vertices are visible from BA^1 and all of them visible from B are also visible from B'. According to Fig. 9, assume that d_1 is the line from BA^1 and $CA(BA^1)$, d_2 is the line from B to $CA(B)$ and d_3 is the line from B' to $CA(B')$. The vertices of $V1 \cup V2 \cup V3$ are put on a convex chain from B' to C in such a way that vertices of $V1$ lie inside α, vertices of $V2$ lie inside β and vertices of $V3$ lie above d_3. As the slope of d_3 is more than d_2 and slope of d_2 is more than d_1 so both α and β are non-empty. Therefore, by considering these constraints, our placement supply edges between vertices of left chain and convex vertices in visibility graph. So we have to consider some constraints to supply edges between right chain and convex vertices. We know that all vertices of the right chain between A and $AC(BA^1)$, are invisible to convex chain (Because $AC(BA^1)$ blocks visibility of B' and BA^1). So, by placing convex vertices above d_1 (includes both regions α and β) these invisibilities can be supplied and we do not need to add more constraints for it. Moreover, Any convex vertex which are visible to B' (equivalently $V1 \cup V2$) sees $CA(B')$ (because all vertices of the right concave chain are above the line through B' and $CA(B')$ and can not block the visibility of them). Therefore, for any convex vertex $p \in V1 \cup V2$, p sees all vertices of vertices of the right chain between $CA(B')$ and $CA(p)$. Moreover, $CA(p)$ blocks the visibility of p and any vertex of vertices of the right chain between A and $CA(p)$. Hence, p must be placed somewhere above the line, d_p, through $CA(p)A^1$ and $CA(p)$ (See Fig. 9). For any $p' \in V1$ (resp. $p' \in V2$), which is closer to C than p, slope of the line through $CA(p')A^1$ and $CA(p')$ is more than d_p and d_1 (resp. d_2) and less than d_2 (resp. d_3). Moreover all these lines will intersect each other above d_3. Consider and arbitrary convex curve from B', that lies completely below d_3 and intersect d_1, d_2 and $d_{CB(B')}$ strictly below d_3. So, By considering all constraints above, we can place all vertices of the set $V1 \cup V2$ on this curve, so that, they supply edges of the induced visibility graph on vertices of the sub-chain from A to $CA(B')$ and vertices of $V1 \cup V2$, and form a convex chain with respect to their order in the Hamiltonian cycle.

The convex vertices from B' to C and the concave vertices from $CA(B')$ to C build a spiral polygon and we have already build a convex sub-polygon from it with boundary vertices $\{B', CA(B')\} \cup V1 \cup V2$ (with B' as one of its joint vertices, and other vertices as all neighbors of B'). So the remaining vertices of this spiral polygon (equivalently, the remaining vertices of the anchor polygon) can be reconstructed according to the method described in Sect. 2 in a close neighborhood of $CA(B')$, so that its boundary do not intersect the rest of the anchor polygon. Note that this reconstruction forces the vertices of $V3$ to be located above d_3. This completes our induction proof for $i = 1$.

Now we can prove the induction step of our proof. From the proof of the base step of the induction, we conclude that for any anchor polygon P, with pair of $G(V, E)$ as its visibility graph, H as its Hamiltonian cycle and A, B and C as its top, left and right joint vertices, respectively, and BC^i as the left joint vertex found by our algorithm, we have another anchor polygon P' such that, its visibility graph and Hamiltonian cycle are isomorphic to $G(V, E)$ and H, respectively, and A, $B' = BC^1$ and C as its top, left and right joint vertices, respectively. So in P' our algorithm will find $B'C^{i-1}$ as its left joint vertex. So, by repeating this step, there will be a polygon P'' with visibility graph and Hamiltonian cycle isomorphic to $G(V, E)$ and H, respectively and A, BC^i and C as its top, left and right joint vertices, respectively. So the theorem is true. $\square$

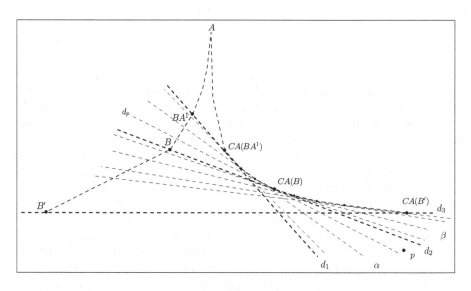

Fig. 9. Lines d_1 and d_2, parts α and β and other constraints

References

1. Everett, H.: Visibility graph recognition, Ph.d. Dissertation, University of Toronto (1990)

2. Everett, H., Corneil, D.G.: Recognizing visibility graphs of spiral polygons. J. Algorithms **11**(1), 1–26 (1990)
3. Colley, P., Lubiw, A., Spinrad, J.P.: Visibility graphs of towers. Comput. Geom. **7**, 161–172 (1997)
4. Cormen, T.H., Leiserson, C.E., Rivest, R.L., Stein, C.: Introduction to Algorithms, 3 edn. MIT Press (2009). ISBN 978-0-262-03384-8, pp. I-XIX, 1–1292
5. Ghosh, S.K.: On recognizing and characterizing visibility graphs of simple polygons. In: Karlsson, R., Lingas, A. (eds.) SWAT 1988. LNCS, vol. 318, pp. 96–104. Springer, Heidelberg (1988)
6. Ghosh, S.K.: On recognizing and characterizing visibility graphs of simple polygons. Discrete Comput. Geom. **17**(2), 143–162 (1997)
7. Streinu, I.: Non-stretchable pseudo-visibility graphs. Comput. Geom. **31**(3), 195–206 (2005)
8. Richter-Gebert, J.: Mnëvs universality theorem revisited. Séminaire Lotaringien de Combinatoire (1995)
9. Ghosh, S.K.: Visibility Algorithms in the Plane. Cambridge University Press (2007). ISBN 0521875749
10. Teller, S.J., Hanrahan, P.: Global visibility algorithms for illumination computations. In: SIGGRAPH, pp. 239–246 (1993)
11. Belta, C., Isler, V., Pappas, G.J.: Discrete abstractions for robot motion planning and control in polygonal environments. IEEE Trans. Robot. **21**(5), 864–874 (2005)
12. Abello, J., Egecioglu, Ö., Kumar, K.: Visibility graphs of staircase polygons and the weak Bruhat order, I: from visibility graphs to maximal chains. Discrete Comput. Geom. **14**(3), 331–358 (1995)
13. Coullard, C.R., Lubiw, A.: Distance visibility graphs. Int. J. Comput. Geom. Appl. **2**(4), 349–362 (1992)
14. Elgindy, H.A.: Hierarchical Decomposition of Polygons with Applications. McGill University, Quebec (1985)
15. Abello, J., Hua, L., Sekhar, P.: On visibility graphs of simple polygons. Congressus Numerantium **90**, 119–128 (1992)
16. Abello, J., Kumar, K.: Visibility graphs of 2-spiral polygons (extended abstract). In: Baeza-Yates, R., Poblete, P.V., Goles, E. (eds.) LATIN 1995. LNCS, vol. 911, pp. 1–15. Springer, Heidelberg (1995)
17. Abello, J., Kumar, K.: Visibility graphs and oriented matroids (extended abstract). In: Tamassia, R., Tollis, I.G. (eds.) GD 1994. LNCS, vol. 894, pp. 147–158. Springer, Heidelberg (1995)

Automating the Verification of Realtime Observers Using Probes and the Modal mu-calculus

Silvano Dal Zilio[1,2]([✉]) and Bernard Berthomieu[1,2]

[1] LAAS, CNRS, 7 avenue colonel Roche, 31400 Toulouse, France
dalzilio@laas.fr
[2] LAAS, Université de Toulouse, 31400 Toulouse, France

Abstract. A classical method for model-checking timed properties—such as those expressed using timed extensions of temporal logic—is to rely on the use of observers. In this context, a major problem is to prove the correctness of observers. Essentially, this boils down to proving that: (1) every trace that contradicts a property can be detected by the observer; but also that (2) the observer is innocuous, meaning that it cannot interfere with the system under observation. In this paper, we describe a method for automatically testing the correctness of realtime observers. This method is obtained by automating an approach often referred to as *visual verification*, in which the correctness of a system is performed by inspecting a graphical representation of its state space. Our approach has been implemented on the tool Tina, a model-checking toolbox for Time Petri Net.

1 Introduction

A classical method for model-checking timed behavioral properties—such as those expressed using timed extensions of temporal logic—is to rely on the use of observers. In this approach, we check that a given property, P, is valid for a system S by checking the behavior of the system composed with an observer for the property. That is, for every property P of interest, we need a pair (Obs_P, ϕ_P) of a system (the observer) and a formula. Then property P is valid if and only if the composition of S with Obs_P, denoted $(\mathsf{S} \,\|\, \mathsf{Obs}_P)$, satisfies ϕ_P. This approach is useful when the properties are complex, for instance when they include realtime constraints or involve arithmetic expressions on variables. Another advantage is that we can often reduce the initial verification problem to a much simpler model-checking problem, for example when ϕ_P is a simple reachability property.

In this context, a major problem is to prove the correctness of observers. Essentially, this boils down to proving that every trace that contradicts a property can be detected. But this also involves proving that an observer will never block the execution of a valid trace; we say that it is *innocuous* or non-intrusive.

This work was partly supported by the ITEA2 Project OpenETCS.

Published by Springer International Publishing Switzerland 2016. All Rights Reserved.
M.T. Hajiaghayi and M.R. Mousavi (Eds.): TTCS 2015, LNCS 9541, pp. 90–104, 2016.
DOI: 10.1007/978-3-319-28678-5_7

In other words, we need to assure that the "measurements" performed by the observer can be made without affecting the system.

In the present work, we propose to use a model-checking tool chain in order to check the correctness of observers. We consider observers related to linear time properties obtained by extending the pattern specification language of Dwyer et al. [7] with hard, realtime constraints. In this paper, we take the example of the pattern "**Present** a **after** b **within** $[d_1, d_2[$", meaning that event a must occur after d_1 units of time (u.t.) of the first occurrence of b, if any, but not later than d_2. Our approach can be used to prove both the soundness and correctness of an observer when we fix the values of the timing constraints (the values of d_1 and d_2 in this particular case).

Our method is not enough, by itself, to prove the correctness of a verification tool. Indeed, to be totally trustworthy, this will require the use of more heavy-duty software verification methods, such as interactive theorem proving. Nonetheless our method is complementary to these approaches. In particular it can be used to debug new or optimized definitions of an observer for a given property before engaging in a more complex formal proof of its correctness.

Our method is obtained by automating an approach often referred to as *visual verification*, in which the correctness of a system is performed by inspecting a graphical representation of its state space. Instead of visual inspection, we check a set of branching time (modal μ-calculus) properties on the discrete time state space of a system. These formulas are derived automatically from a definition of the pattern expressed as a first-order formula over timed traces. The gist of this method is that, in a discrete time setting, first-order formulas over timed traces can be expressed, interchangeably, as regular expressions, LTL formulas or modal μ-calculus formulas.

This approach has been implemented on the tool Tina [4], a model-checking toolbox for Time Petri Net [11] (TPN). This implementation takes advantage of several components of Tina: state space exploration algorithms with a discrete time semantics (using the option -F1 of Tina); model-checkers for LTL and for modal μ-calculus, called *selt* and *muse* respectively; a new notion of *verification probes* recently added to Fiacre [3,5], one of the input specification language of Tina. While model checkers are used to replace visual verification, probes are used to ensure innocuousness of the observers.

Outline and Contributions. The rest of the paper is organized as follows. In Sect. 2, we give a brief definition of Fiacre and the use of probes and observers in this language. In Sect. 3, we introduce the technical notations necessary to define the semantics of patterns and timed traces and focus on an example of timed patterns. Before concluding, we describe the graphical verification method and show how to use a model-checker to automate the verification process[1].

The theory and technologies underlying our verification method are not new: model-checking algorithms, semantics of realtime patterns, connection between path properties and modal logics, ... Nonetheless, we propose a novel way to combine these techniques in order to check the implementation of observers and

---

[1] Code is available at http://www.laas.fr/fiacre/examples/visualverif.html.

in order to replace traditional "visual" verification methods that are prone to human errors.

Our paper also makes some contributions at the technical level. In particular, this is the first paper that documents the notion of probe, that was only recently added to Fiacre. We believe that our (language-level) notion of probes is interesting in its own right and could be adopted in other specification languages.

2   The Fiacre Language

We consider systems modeled using the specification language Fiacre [3,5]. (Both the system and the observers are expressed in the same language.) Fiacre is a high-level, formal specification language designed to represent both the behavioral and timing aspects of reactive systems.

Fiacre programs are stratified in two main notions: *processes*, which are well-suited for modeling structured activities, and *components*, which describes a system as a composition of processes. Components can be hierarchically composed. We give in Fig. 1 a simple example of Fiacre specification for a computer mouse button capable of emitting a double-click event. The behavior, in this case, is to emit the event double if there are more than two click events in strictly less than one unit of time (u.t.).

```
process Push [click   : none,
              single  : none,
              double  : none,
              delay   : none] is

states s0, s1, s2

var dbl : bool := false

from s0 click; to s1

from s1
  select
     click; dbl := true; loop
  [] delay; to s2
  end

from s2
  if dbl then double
  else single end;
  dbl := false; to s0
```

```
component Mouse [click   : none,
                 single  : none,
                 double  : none] is

port delay : none in [1,1]

priority delay > click

par
    Push [click, single, double, delay]
end
```

Fig. 1. A double-click example in Fiacre

Processes. A process is defined by a set of parameters and control states, each associated with a set of *complex transitions* (introduced by the keyword **from**). The initial state of a process is the state corresponding to the first **from** declaration.

Complex transitions are expressions that declare how variables are updated and which transitions may fire. They are built from deterministic constructs available in classical programming languages (assignments, conditionals, sequential composition, . . .); non-deterministic constructs (such as external choice, with the **select** operator); communication on ports; and jump to a state (with the **to** or **loop** operators).

For example, in Fig. 1, we declare a process named Push with four communication ports (click to delay) and one local boolean variable, dbl. Ports may send and receive typed data. The port type none means that no data is exchanged; these ports simply act as synchronization events. Regarding complex transitions, the expression related to state s1 of Push, for instance, declares two possible transitions from s1: (1) on a click event, set dbl to true and stay in state s1; and (2) on a delay event, change to state s2.

Components. A component is built from the parallel composition of processes and/or other components, expressed with the operator **par** P_0 || . . . || P_n **end**. In a composition, processes can interact both through synchronization (message-passing) and access to shared variables (shared memory).

Components are the unit for process instantiation and for declaring ports and shared variables. The syntax of components allows to associate timing constraints with communications and to define priorities between communication events. The ability to express directly timing constraints in programs is a distinguishing feature of Fiacre. For example, in the declaration of component Mouse (see Fig. 1), the **port** statement declares a local event delay and asserts that a transition from s1 to s2 should take exactly one unit of time. (Time passes at the same rate for all the processes.) Additionally, the **priority** statement asserts that a transition on event click cannot occur if a transition on delay is also possible.

Probes and Observers. The Fiacre language has been extended, recently, to allow the definition of observers, which are a distinguished category of subprograms that interact with other Fiacre components only through the use of *probes*. A probe is used to observe modifications in the system without interfering with it; probes react to the occurrence of an event without engaging in it.

A typical probe declaration is of the form path/obs, where obs denotes the observable and path defines its context, that is a path to the component (or process) instance where obs is defined (see for example http://www.laas.fr/fiacre/properties.html). In our setting, observable events are instantaneous actions involved in the evolution of the system: it can be a synchronization over a port p (denoted event p); a process that enters the state s (denoted state s); or an expression including shared variables, say exp, that changes value (denoted value exp). For instance, in the case of the Mouse component of Fig. 1, a probe triggered when the (only instance of) process Push is in state s2 would have the form (Mouse/1/state s2).

The use of probes greatly simplifies the proof of innocuousness of an observer. In particular, with probes, an observer can only influence a system by "blocking the evolution of time", that is by performing an infinite sequence of actions in

```
process NeverTwice [a:sync] is        component Obs is
  states idle, once, error              port p:sync is Mouse/event click
  from idle a ; to once                 par NeverTwice [p] end
  from once a ; to error
```

Fig. 2. A simple observer example

finite time. Therefore, proving that an observer is innocuous amounts to proving that it has no Zeno behaviors, which is always possible when a system is bounded.

An observer is a Fiacre component where ports are associated to probes (using the keyword **is**); ports associated with a probe have the reserved type sync. We give a naive example of observer in Fig. 2, where the component Obs monitors synchronizations on the event click. In this example, the process neverTwice will reach the state error if its probe parameter, a, is triggered more than once.

In the remainder of the text, we use the notation (Mouse || Obs) to denote the program obtained by concatenating the declaration of these two components (i.e. the code from Fig. 1 with the code from Fig. 2). As a consequence, we are able to detect if the system can emit two single click events just by checking if the process neverTwice can reach the state error in (Mouse || Obs).

3   Timed Traces and First-Order Formulas Over Traces

The semantics of Fiacre (and the properties we want to check) are based on a notion of *timed traces*, which are sequences mixing events and time delays. In this context, a "realtime property" can be defined as a set of timed traces, which define timing and behavioral constraints on the acceptable execution of a system. In this work, we consider properties derived from realtime patterns, that can be expressed using first-order formulas over timed traces.

Timed Traces. In our context, observable events are: communication on a port; the change of state of a process; and the change of value of a variable. We use a dense time model, meaning that we consider rational time delays and work both with strict and non-strict time bounds. Hence a timed trace is a (possibly infinite) sequence of events $a, b, \ldots$ and durations $\delta \in \mathbb{Q}^+$:

$$\sigma ::= \epsilon \mid \sigma a \mid \sigma \delta$$

Given a finite trace σ and a—possibly infinite—trace σ', we denote $\sigma\sigma'$ the *concatenation* of σ and σ'. We will also use the expression $\Delta(\sigma)$ to denote the duration (time length) of a trace σ, that is the sum of the individual delays in σ. The semantics of a system expressed with Fiacre, say S, can be defined as a set $[\![S]\!]$ of timed traces. We use the notation $\sigma \models S$ when the trace σ is in the set $[\![S]\!]$. The semantics of a property (timed pattern) will be expressed as the set of all timed traces where the pattern holds. We say that a system S satisfies a timed requirement P if $[\![S]\!] \subseteq [\![P]\!]$.

Realtime Properties and Their Semantics. We propose to define properties using First-Order Formulas over Timed Traces (FOTT). A FOTT formula $\Phi(x)$, with free variables $x = (x_1, \ldots, x_n)$, is a first-order logic formula over traces with equality between traces ($\sigma = \sigma'$), comparison between a duration and an interval ($\Delta(\sigma) \in I$) and concatenation ($\sigma = \sigma_1 \sigma_2$).

$$\Phi(x) ::= \Phi \wedge \Phi' \mid \neg \Phi \mid \exists x . \Phi \mid (x = \sigma) \mid (x = y \, z) \mid (\Delta(x) \in I)$$

For instance, when referring to a timed trace σ and an event a, the following formula is a tautology if the event a does not occur in σ:

$$(a \notin \sigma) \stackrel{\text{def}}{=} \neg (\exists x_1, x_2, x_3 . (\sigma = x_1 \, x_2) \wedge (x_2 = a \, x_3))$$

Likewise, we can define the "scope" σ **after** b—that determines the part of a trace σ located after the first occurrence of b—as the trace σ' denoted by the first-order formula: $\exists x, y . (\sigma = x \, y) \wedge (y = b \, \sigma') \wedge (b \notin x)$.

The semantics of a formula $\Phi(x_1, \ldots, x_n)$ is a set of valuation functions ς associating a trace $\sigma_i = \varsigma(x_i)$ to each of the variables x_i with $i \in 1..n$, also denoted $[x_i \mapsto \sigma_i]_{i \in 1..n}$. The semantics of Φ can be defined inductively as follows:

$$\llbracket \Phi(x) \wedge \Psi(x) \rrbracket = \llbracket \Phi(x) \rrbracket \cap \llbracket \Psi(x) \rrbracket \qquad \llbracket x = \sigma \rrbracket = \{\varsigma \mid \varsigma(x) = \sigma\}$$
$$\llbracket \exists y . \Phi(x) \rrbracket = \{\varsigma \mid \varsigma + [y \mapsto \sigma] \in \llbracket \Phi(x) \rrbracket\} \quad \llbracket x = y \, z \rrbracket = \{\varsigma \mid \varsigma(x) = \varsigma(y) \, \varsigma(z)\}$$
$$\llbracket \Delta(x) \in I \rrbracket = \{\varsigma \mid \Delta(\varsigma(x)) \in I\}$$

With these definitions, a *regular set of timed traces* is the set of traces "solutions" of an existential FOTT formula with a single free variable, $\Phi(x)$; that is the set of traces σ such that the valuation $[x \mapsto \sigma]$ is in $\llbracket \Phi(x) \rrbracket$.

In this paper, we will mainly restrict ourselves to the special case of timed traces where events occur at integer dates; i.e. we restrict delays δ to be in $\mathbb{N}$ rather than in $\mathbb{Q}^+$. These traces can be generated using a "discrete time" abstraction of the models, where special transitions (labeled with t) are used to model the flow of time. Label t stands for the "tick" of the logical clock.

The discrete time semantics will be enough to prove all the properties needed in our study. Indeed, when a model contains only "closed timing constraints" (of the kind $[d_1, d_2]$ or $[d_1, \infty[$), the discrete time semantics is enough to check reachability properties.

With discrete time, a delay δ can be replaced by sequences of δ t's, and therefore a finite timed trace can be simply interpreted as a word. In the remainder, we also consider a special symbol, z, that stands for internal actions of the system. Hence it is possible to interpret the semantics of (discrete) FOTT specification as a language over the alphabet $A = \{z, t, a, b, \ldots\}$. Actually, in the discrete case, we can show that a regular set of timed traces is also a regular language. For example, the semantics of the formula $\exists y, z, w . ((x = y \, z) \wedge (z = a \, w))$ is the regular language corresponding to the expression $A* \cdot a \cdot A*$.

This connection between different type of logics is at the core of our approach. Our method could be applied to more high-level property languages, such as timed extension of temporal logic [10], but would require a more complex encoding into LTL when modalities can be nested.

Our Running Example: The Present Pattern. Users of Fiacre have access to a catalog of specification patterns based on a hierarchical classification borrowed from Dwyer [7]. Patterns are built from five basic categories—existence, absence, universality, response and precedence—and can be composed using logical connectives. In each category, generic patterns may be specialized using *scope modifiers*—such as before, after, between—that limit the range of the execution trace over which the pattern must hold. Finally, timed patterns are obtained using one of two possible kinds of *timing modifiers* that limit the possible dates of events referred in the pattern: **within** I—used to constrain the delay between two given events to be in the time interval I—and **lasting** d—used to constrain the length of time during which a given condition holds (without interruption) to be greater than d.

Due to limited space, we study only one example of timed pattern, namely **Present** a **after** b **within** $[d_1, d_2[$. A complete catalog is available in [1]. This is a simple example of existence patterns. Existence patterns are used to express that, in every trace of the system, some events must occur. This pattern holds for traces such that the event a occurs at a date t_0 after the first occurrence of b with $t_0 \in [d_1, d_2[$. The property is also satisfied if b never holds. Hence traces σ that satisfy this pattern are models of the existential FOTT formula:

$$\text{Pres}(x) \stackrel{\text{def}}{=} (b \notin x) \ \vee \ \exists y, z, w \ . \ ((x = y \, b \, z \, a \, w) \wedge (b \notin y) \wedge (\Delta(z) \in [d_1, d_2[))$$

```
process Present [a:sync, b:sync] is
    states idle, start, watch, error, stop
    from idle  b;  to start
    from start wait [d₁, d₁]; to watch
    from watch select
            a; to stop
            unless
                wait [d₂ - d₁, ···[; to error
            end
```

Listing 1.1. Observer for the pattern: **Present** a **after** b **within** $[d_1, d_2[$

With the discrete semantics, formula $\text{Pres}(x)$ matches exactly the words of the form $w_1 \, b \, w_2 \, a \, w_3$ where w_1 contains no occurrences of b and w_2 contains exactly k occurrences of t with $k \in [d_1, d_2[$. (This is a regular language.) We show in the next section how to (semi-)automatically generate the regular expression corresponding to such FOTT formulas.

We give an example of observer associated to this pattern in Listing 1.1. This observer is composed of one process that monitors the system through the ports a and b (that should be instantiated with the relevant probes). The process is initially in state idle and moves to start when b is triggered. When in state start for d_1 unit of time, the observer moves to state watch (this is the meaning of the **wait** operator). The **select** operator is a non-deterministic choice, with

unless coding priorities. Hence, in state watch, the observer moves to stop if an a occurs, unless a duration equals to $(d_2 - d_1)$ elapses, in which case it moves to the state error. As a consequence, the pattern is false whenever the probe (Present/state error) is reachable. Hence the formula associated to the pattern is $\phi_P \stackrel{\text{def}}{=} [] - (\text{Present/state error})$.

To prove that an observer Obs for the pattern P is correct, we need to prove that, for every system S, the program (S ‖ Obs) satisfies the formula ϕ_P if and only if $[\![S]\!] \subseteq [\![P]\!]$. In [1], we have defined a mathematical framework to formally prove these kind of properties, but this framework relies on manual proofs and is not supported by any tooling. Efforts are also under way to completely mechanize these proofs using the Coq proof assistant [8]. Nonetheless, formal proofs of correctness can be quite tedious. Therefore, to detect possible problems with an observer early on (that is, before spending a lot of efforts doing a formal proof of correctness) we also rely on a "visual" verification method, that is akin to debugging our observers.

In the next section, we show how to apply the visual verification approach on our running example. One of the objectives of our work is to replace this visual verification step with a more formal approach. This is done in Sect. 5.

4 Visual Verification of Observers

In the remainder of this section, we describe the visual verification method using the particular case of the pattern **Present** a **after** b **within** [4, 5[; we assume that Obs is the observer Present defined in Listing 1.1, that $d_1 = 4$ and that $d_2 = 5$.

To prove that the observer Present is correct, we need to prove, for every system S, the equivalence between two facts: (1) the state (Present/state error) is not reachable in the program (S ‖ Present[a, b]); and (2) the traces of S are valid for the property Pres, i.e. $[\![S]\!] \subseteq [\![\text{Pres}]\!]$.

The first step is to get rid of the universal quantification on all possible systems, S, that is introduced by our definition of correctness. The idea is to check the observer on a particular Fiacre program—called Universal—that can generate all possible combinations of delays and events a, b and z. We give an example of universal process in Listing 1.2. The process Universal has only one state and three possible transitions. Each transition changes the value of a shared integer variable, x. The first and second transitions of Universal can be fired without time constraints. In our context, the probe a will be triggered to the event "setting x to 1" and b to "setting x to 2". The third transition resets the value of x to 0 immediately and corresponds to the internal event z.

We can now use our verification toolchain to generate the state graph for the program (Universal ‖ Present) using a discrete time exploration construction. This can be obtained using the flag −F1 in Tina (it is possible to generate a state graph with many different abstractions with Tina, including dense time models).

The resulting graph is displayed in Fig. 3. This state graph has been generated and printed using the tool *nd*, which is also part of the Tina toolset; nd is an editor and animator for extended Time Petri Nets that can export nets and state

```
process Universal (&x : nat) is
   states s0
   from s0 select
        x := 1; to s0        /* setting x to 1 */
        [] x := 2; to s0      /* setting x to 2 */
      unless
        on (x <> 0); wait [0,0]; x := 0; to s0
      end

component Main is
   var x : nat := 0
   port a : sync is value (x = 1),  b : sync is value (x = 2)
   par Universal (&x) || Present [a, b] end
```

Listing 1.2. Universal program in Fiacre

graphs in several, machine readable formats. This graph has only 26 states and can therefore be easily managed manually. The main factor commanding the number of states is the value of the timing constraints used in the pattern; in our observations, all the generated state graphs were of manageable size.

The transitions in the state graph are also quite straightforward: we find the visible and internal transitions as before, labeled with a, b, z and t. For ease of reading, we have also changed the labels of internal transitions in the observer Present. For instance, the transition from state 2 to 3 corresponds to the observer entering the state start; likewise for the transitions labeled with watch, stop and error. The states where the observer is in state error (the states that contradict the property $\phi_P \overset{\text{def}}{=} []$ - (Present/state error)) are $Errors = \{20, 22, 23\}$.

We can already debug the pattern **Present** a **after** b **within** $[4, 5[$ by visually inspecting the state graph.

For *soundness*, we need to check that, when the pattern is not satisfied—for traces σ that do not satisfy formula Pres—then the observer will detect a problem (observer Present eventually reaches a state in the set $Errors$).

For *innocuousness* we need to check that, from any state, it is always possible to reach a state where event a (respectively b and t) can fire. Indeed, this means that the observer cannot selectively remove the observation of a particular sequence of external transitions or the passing of time.

This graphical verification method has some drawbacks. As such, it relies on a discrete time model and only works for fixed values of the timing parameters (we have to fix the value of d_1 and d_2). Nonetheless, it is usually enough to catch many errors in the observer before we try to prove the observer correct more formally.

5 Automating the Visual Verification Method

A problem with the previous approach is that it essentially relies on an informal inspection (and on human interaction). We show how to solve this problem by

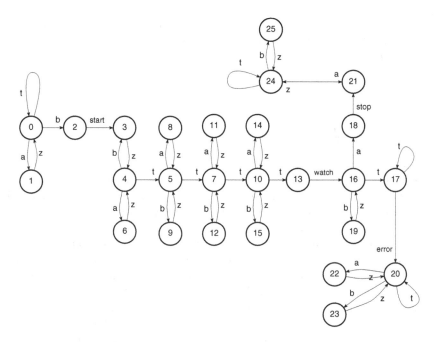

Fig. 3. State graph for (Universal || Present)

replacing the visual inspection of the state graph by the verification of modal μ-calculus formulas. (the Tina toolset includes a model-checker for the μ-calculus called *muse*.) The general idea rests on the fact that we can interpret the state graph as a finite state automaton and (some) sets of traces as regular languages. This analogy is generally quite useful when dealing with model-checking problems. We start by defining some useful notations.

Label Expressions are boolean expressions denoting a set of (transition) labels. For instance, $A_{ext} = (a \vee b)$ denotes the external transitions, while the expression $-(a \vee b \vee t)$ is only matched by the silent transition label. We will also use the expression $\top$ to denote the disjunction of all possible labels, e.g. $\top = (-b) \vee b$. The model checker *muse* allows the definition of label expressions using the same syntax.

Regular (Path) Expressions. In the following, we consider regular expressions built from label expressions. For example, the regular expression $t \cdot (-t)*$ denotes traces of duration 1 with no events occurring at time 0.

$$Tick \stackrel{\text{def}}{=} t \cdot (-t)* \tag{1}$$

We remark that it is possible to define the set of (discrete) traces where the FOTT formula Pres holds using the union of two regular languages: (1) the traces where b never occurs, (-b)*; and (2) the traces where there is an a four units of time after the first b. The latter corresponds to the regular expression $(x = y \, \mathsf{b} \, z \, \mathsf{a} \, w) \wedge (\mathsf{b} \notin y) \wedge (\Delta(z) \in [4, 5[)$

$$Pres \stackrel{\text{def}}{=} R_1 \vee R_2 \tag{2}$$

$$R_1 \stackrel{\text{def}}{=} (\text{-b})^* \tag{3}$$

$$R_2 \stackrel{\text{def}}{=} (\text{-b})^* \cdot \mathsf{b} \cdot (\text{-t})^* \cdot Tick \cdot Tick \cdot Tick \cdot Tick \cdot \mathsf{a} \cdot \top^* \tag{4}$$

By construction, the regular language associated to $R_1 \vee R_2$ is exactly the set of finite traces matching (the discrete semantics) of Pres. In the most general case, a regular expression can always be automatically generated from an existential FOTT formula when the time constraints of delay expressions are fixed (the intervals I in the occurrences of $(\Delta(x) \in I)$).

The next step is to check that the observer agrees with every trace conforming to R_2. For this we simply need to check that, starting from the initial state of (Universal $\|$ Present), it is not possible to reach a state in the set Errors by following a sequence of transitions labeled by a word in R_2.

This is a simple instance of a language inclusion problem between finite state automata. More precisely, if Present is the set of states visited when accepting the traces in $R_1 \vee R_2$, we need to check that Errors is included in the complement of the set Present (denoted $\overline{Present}$). In our example of Fig. 3, we have that $\overline{Present} = \{17, 20, 22, 23\}$, and therefore Errors $\subseteq \overline{Present}$.

This automata-based approach has still some drawbacks. This is what will motivate our use of a branching time logic in the next section. In particular, this method is not enough to check the soundness or the innocuousness of the observer. For innocuousness, we need to check that every event may always eventually happen. Concerning soundness, we need to prove that Errors $\supseteq \overline{Present}$; which is false in our case. The problem lies in the treatment of time divergence (and of fairness), as can be seen from one of the counter-example produced when we use our LTL model-checker to check the soundness property, namely: b.start.z.t.t.t.t.watch.t.t.··· (ending with a cycle of t transitions). This is an example where the error transition is continuously enabled but never fired.

Branching Time Specification. We show how to interpret regular expressions over traces using a modal logic. In this case, the target logic is a modal μ-calculus with operators for forward and backward traversal of a state graph . (Many temporal logics can be encoded in the μ-calculus, including CTL*). In this context, the semantics of a formula ψ over a Kripke structure (a state graph) is the set of states where ψ holds.

$$\psi ::= \phi \wedge \psi \mid \neg \psi \mid <A> \psi \mid \psi <A> \mid X \mid (\min X \mid \psi)$$

The basic modalities in the logic are $<A>\psi$ and $\psi<A>$, where A is a label expression. A state s is in $<A>\psi$ if and only if there is a (successor) state s' in ψ and a transition from s to s' with a label in A. Symmetrically, s is in $\psi<A>$

if and only if there is a (predecessor) state s' in ψ and a transition from s' to s with a label in A. In the following, we will also use two constants, T, the true formula (matching all the states), and '0, that denotes the initial state of the model; and the least fixpoint operator min X | ψ(X).

For example, the formula <a>T matches all the states that are the source of an a-transition, likewise Reach_a $\overset{\text{def}}{=}$min X | (<a>T $\vee$ <Z>X) matches all the states that can lead to an a-transition using only internal transitions. As a consequence, we can test innocuousness by checking that the formula (Reach_a $\wedge$ Reach_b $\wedge$ Reach_t) is true for all states.

The soundness proof relies on an encoding from regular path expressions into modal formulas. We define two encodings: $((R))$ that matches the states encountered while firing a trace matching a regular expression R; and $((R))_e$ that matches the state reached (at the end) of a finite trace in R. These encodings rely on two derived operators. (Again, we assume here that A is a label expression.)

$$\psi \circ A \overset{\text{def}}{=} \psi\texttt{<A>} \qquad\qquad \psi * A \overset{\text{def}}{=} \texttt{min X } | \ \psi \vee X\texttt{<A>}$$

$((R \cdot A))_e$	$\overset{\text{def}}{=} ((R))_e \circ A$	$((R \cdot A))$	$\overset{\text{def}}{=} ((R)) \vee ((R \cdot A))_e$
$((R \cdot A^*))_e$	$\overset{\text{def}}{=} ((R))_e * A$	$((R \cdot A^*))$	$\overset{\text{def}}{=} ((R)) \vee ((R \cdot A^*))_e$
$((R \cdot Tick))_e$	$\overset{\text{def}}{=} (((R))_e \circ t) * (-t)$	$((R \cdot Tick))$	$\overset{\text{def}}{=} ((R)) \vee ((R \cdot Tick))_e$
$((R_1 \vee R_2))_e$	$\overset{\text{def}}{=} ((R_1))_e \vee ((R_2))_e$	$((R_1 \vee R_2))$	$\overset{\text{def}}{=} ((R_1)) \vee ((R_2))$
$((\epsilon))_e$	$\overset{\text{def}}{=} \ '0$	$((\epsilon))$	$\overset{\text{def}}{=} \ '0$

Lemma 1. *Given a Kripke structure K, the states matching the formula $((R))_e$ (respectively $((R))$) in K are the states reachable from the initial state after firing (resp. all the states reachable while firing) a sequence of transitions matching R.*

Proof (Sketch). By induction on the definition of R. For example, if we assume that ψ correspond to the regular expression R, then $\psi * A$ matches all the states reachable from states where ψ is true using (finite) sequences of transition with label in A; i.e. formula $\psi * A$ corresponds to $R \cdot A^*$. Likewise, we use the interpretation of the empty expression, ϵ, to prefix every formula with the constant '0 (that will only match the initial state). This is necessary since μ-calculus formulas are evaluated on all states whereas regular path expressions are evaluated from the initial state.   □

For example, we give the formula for $((R_2))_e$ below, where $\psi \circ \texttt{Tick}$ stands for the expression $(\psi \circ \texttt{t}) * (\texttt{-t})$:

$$((R_2))_e \overset{\text{def}}{=} \ '0 * (\texttt{-b}) \circ \texttt{b} * (\texttt{-t}) \circ \texttt{Tick} \circ \texttt{Tick} \circ \texttt{Tick} \circ \texttt{Tick} \circ \texttt{a} * \texttt{T}$$

If ψ_{Err} is a modal μ-calculus formula that matches the error condition of the observer, then we can check the correctness and soundness of the observer Present by proving that the equivalence (EQ), below, is a tautology (that it is true on every states of (Universal || Present)).

$$((Pres)) \ \Leftrightarrow \ -\psi_{Err} \tag{EQ}$$

Again, we can interpret the "error condition" using the μ-calculus. The definition of errors is a little bit more involved than in the previous case. We say that a state is in error if the transition error is enabled (the formula `<error>T` is true) or if the state can only be reached by firing the error transition (which corresponds to the formula `(T<error> * T)` $\wedge$ `('0 * (- error))`). Hence ψ_{Err} is the disjunction of these two properties:

$$\psi_{Err} \overset{\text{def}}{=} \texttt{<error>T} \ \vee \ ((\texttt{T<error>} * \texttt{T}) \ \wedge \ -(\texttt{'0} * (-\texttt{error})))$$

The formula (EQ) can be checked almost immediately (less than 1 s on a standard computer) for models of a few thousands states using *muse*. Listing 1.3 gives a *muse* script file that can be used to test this equivalence relation.

```
# Results are displayed as set of states. Use "output card" to see the cardinality
output set;

# definition of derived operators

infix X * L = min Y | X ∨ Y⟨L⟩;      infix X o L = X⟨L⟩;
op TICK X = min Y | X⟨t⟩ ∨ Y⟨−t⟩;   op NEVER L = ('0) * (−L);
op EXT = a ∨ b ∨ t; # labels of the external transitions
op REACH L = min X | (⟨L⟩T) ∨ ⟨−EXT⟩X;

# INNOCUOUSNESS

op Innocuous = (REACH a) ∧ (REACH b) ∧ (REACH t);

# SOUNDNESS

op A0 = (NEVER b) o b;        op S0 = (NEVER b) ∨ A0;
op A1 = A * (−t);            op S1 = S0 ∨ A1;
op A2 = TICK(A1);           op S2 = S1 ∨ A2;
op A3 = TICK(A2);           op S3 = S2 ∨ A3;
op A4 = TICK(A3);           op S4 = S3 ∨ A4;
op A5 = TICK(A4);           op S5 = S4 ∨ A5;
op A6 = A5 o a;             op S6 = S5 ∨ A6;
op A7 = A6 * T;             op S7 = S6 ∨ A7;

op R1 = NEVER b;            op R2 = S7
op Pres = R1 ∨ R2;
op ERRORS = ⟨error⟩T ∨ (((T⟨error⟩) * T) ∧ − (('0) * (−error)));

Pres ⇔ (− ERRORS); # this is a tautology if all the states are listed
```

Listing 1.3. Script file for *muse* to check that ((Pres)) $\Leftrightarrow$ $-\psi_{Err}$ is a tautology

6   Related Work and Conclusion

Few works consider the verification of model-checking tools. Indeed, most of the existing approaches concentrate on the verification of the model-checking algorithms, rather than on the verification of the tools themselves. For example, Smaus et al. [15] provide a formal proof of an algorithm for generating Büchi automata from a LTL formula using the Isabelle interactive theorem prover. This algorithm is at the heart of many LTL model checkers based on an automata-theoretic approach. The problem of verifying verification tools also appears in conjunction with certification issues. In particular, many certification norms, such as the DO-178B, requires that any tool used for the development of a critical equipment be qualified at the same level of criticality than the equipment. (Of course, certification does not necessarily mean formal proof!) In this context, we can cite the work done on the certification of the SCADE compiler [14], a tool-suite based on the synchronous language Lustre that integrates a model-checking engine. Nonetheless, only the code-generation part of the compiler is certified and not the verification part. Finally, another possibility is to rely on a kind of "Proof-Carrying Code" approach, where the model checker can produce a deductive proof on either success or failure [12]. This proof can then be checked separately, using a tool independent from the model checker.

Concerning observer-based model-checking, most of the works rely on an automatic way to synthesize observers from a formal definition of the properties. For instance, Aceto et al. [2] propose a method to verify properties based on the use of test automata. In this framework, verification is limited to safety and bounded liveness properties since the authors focus on properties that can be reduced to reachability checking. In the context of Time Petri Net, Toussaint et al. [16] also propose a verification technique based on "timed observers", but they only consider four specific kinds of time constraints. None of these works consider the complexity or the correctness of the verification problem. Another related work is [9], where the authors define observers based on Timed Automata for each pattern. Our approach is quite orthogonal to the "synthesis approach". Indeed we seek, for each property, to come up with the best possible observer in practice. To this end, using our toolchain, we compare the complexity of different implementations on a fixed set of representative examples and for a specific set of properties and kept the best candidates. The need to check multiple implementations for the same patterns has motivated the need to develop a lightweight verification method for checking their correctness.

Compared to these works, we make several contributions. We define a complete verification framework for checking observers with hard realtime constraints. This framework has been tested on a set of observers derived from high-level timed specification patterns. This work is also our first public application of the probe technology, that was added to Fiacre only recently. To the best of our knowledge, the notion of *probes* is totally new in the context of formal specification language. Paun and Chechik propose a somewhat similar mechanism in [6,13]—in an untimed setting—where they define new categories of events. However our approach is more general, as we define probes for a richer set of events, such as variables changing state. We believe that this (language-level) notion of probes is interesting in its own right and could be adopted by other formal specification languages. Finally, we propose a formal approach that can be used to gain confidence on the implementation of our model-checking tools and that replaces traditional "visual verification methods" that are prone to human errors.

References

1. Abid, N., Dal Zilio, S., Le Botlan, D.: A formal framework to specify and verify real-time properties on critical systems. Int. J. Crit. Comput.-Based Syst. 5(1/2), 4–30 (2014)
2. Aceto, L., Burgueño, A., Larsen, K.G.: Model checking via reachability testing for timed automata. In: Steffen, B. (ed.) TACAS 1998. LNCS, vol. 1384, pp. 263–280. Springer, Heidelberg (1998)
3. Berthomieu, B., Bodeveix, J.-P., Fillali, M., Hubert, G., Lang, F., Peres, F., Saad, R., Jan, S., Vernadat, F.: The syntax and semantics of fiacre - version 3.0 (2012). http://www.laas.fr/fiacre/
4. Berthomieu, B., Ribet, P.-O., Vernadat, F.: The tool Tina - construction of abstract state spaces for Petri nets and time Petri nets. Int. J. Prod. Res. 42, 14 (2004)
5. Berthomieu, B., Bodeveix, J.-P., Farail, P., Filali, M., Garavel, H., Gaufillet, P., Lang, F., Vernadat, F.: Fiacre: an intermediate language for model verification in the topcased environment. In: Proceedings of ERTS (2008)
6. Chechik, M., Paun, D.O.: Events in property patterns. In: Dams, D.R., Gerth, R., Leue, S., Massink, M. (eds.) SPIN 1999. LNCS, vol. 1680, pp. 154–167. Springer, Heidelberg (1999)
7. Dwyer, M.B., Dillon, L.: Online repository of specification patterns. http://patterns.projects.cis.ksu.edu/
8. Garnacho, M., Bodeveix, J.-P., Filali-Amine, M.: A mechanized semantic framework for real-time systems. In: Braberman, V., Fribourg, L. (eds.) FORMATS 2013. LNCS, vol. 8053, pp. 106–120. Springer, Heidelberg (2013)
9. Gruhn, V., Laue, R.: Patterns for timed property specifications. Electr. Notes Theor. Comput. Sci. 153(2), 117–133 (2006)
10. Koymans, R.: Specifying realtime properties with metric temporal logic. Realtime Syst. 2, 255–299 (1990)
11. Merlin, P.M.: A study of the recoverability of computing systems. Ph.D. thesis (1974)
12. Namjoshi, K.S.: Certifying model checkers. In: Berry, G., Comon, H., Finkel, A. (eds.) CAV 2001. LNCS, vol. 2102, pp. 2–13. Springer, Heidelberg (2001)
13. Paun, D.O., Chechik, M.: Events in events in linear-time properties. CoRR J. vol. cs.SE/9906031 (1999)
14. Esterel technologies. SCADE Tool Suite. http://www.esterel-technologies.com/products/scade-suite
15. Schimpf, A., Merz, S., Smaus, J.-G.: Construction of Büchi automata for LTL model checking verified in Isabelle/HOL. In: Berghofer, S., Nipkow, T., Urban, C., Wenzel, M. (eds.) TPHOLs 2009. LNCS, vol. 5674, pp. 424–439. Springer, Heidelberg (2009)
16. Toussaint, J., Simonot-Lion, F., Thomesse, J.-P.: Time constraints verification methods based on time Petri nets. In: Proceedings of FTDCS. IEEE (1997)

Minimizing Walking Length in Map Matching

Amin Gheibi$^{(\boxtimes)}$, Anil Maheshwari, and Jörg-Rüdiger Sack

School of Computer Science, Carleton University, Ottawa, ON, Canada
{agheibi,anil,sack}@scs.carleton.ca

Abstract. In this paper, we propose a geometric algorithm for a map matching problem. More specifically, we are given a planar graph, H, with a straight-line embedding in a plane, a directed polygonal curve, T, and a distance value $\varepsilon > 0$. The task is to find a path, P, in H, and a parameterization of T, that minimize the sum of the length of walks on T and P whereby the distance between the entities moving along P and T is at most ε, at any time during the walks. It is allowed to walk forwards and backwards on T and edges of H. We propose an algorithm with $\mathcal{O}(mn(m+n)\log(mn))$ time complexity and $\mathcal{O}(mn(m+n))$ space complexity, where m (n, respectively) is the number of edges of H (of T, respectively). As we show, the algorithm can be generalized to work also for weighted non-planar graphs within the same time and space complexities.

1 Introduction

Trajectory data are often obtained from global positioning system (GPS) devices. Such devices have accuracy limitations due to noise, sampling intervals, or poor signals (e.g., inside buildings) thus raw spatial trajectories tend not to be accurate. Under the assumption that the travel captured by the trajectory was following edges of a map (stored as a graph) the map matching problem arises. It asks to find a path on the map that "corresponds well" to the given trajectory. Map matching arises in different contexts and is a necessary step in preprocessing raw data before data mining [1]. A variety of approaches have been used to solve the map matching problem (e.g. geometric, probabilistic methods, fuzzy logic, neural networks). In [2], Chen et al. discussed recent map matching algorithms when a trajectory is obtained from low-frequency GPS data of vehicles driving on a road network. Ruan et al. [3] studied indoor map matching technology based on personal motion states. In [4], Asakura et al. proposed a pedestrian-oriented map matching algorithm in the context of disasters. In this context, refugees have battery-driven mobile GPS terminals and move to shelters at walking speed. They stated that in order to reduce battery consumption (which is vital in this context), they chose a geometric approach in which computation resources are less utilized when compared e.g., with probabilistic methods.

A. Gheibi, A. Maheshwari and J.-R. Sack—Research supported by Natural Sciences and Engineering Research Council of Canada.

© IFIP International Federation for Information Processing 2016
Published by Springer International Publishing Switzerland 2016. All Rights Reserved.
M.T. Hajiaghayi and M.R. Mousavi (Eds.): TTCS 2015, LNCS 9541, pp. 105–120, 2016.
DOI: 10.1007/978-3-319-28678-5_8

In this paper, we focus on geometric approaches. We assume that a map is given as a planar graph via a straight-line embedding in a plane. Therefore, a path in the graph corresponds to a polygonal curve in the plane. A trajectory is given as a directed polygonal curve from a starting point to an ending point. The objective is to find a path in the map which is most similar to the given trajectory.

To measure similarity, [1, 7] observe that methods which consider global features of the input trajectories achieve more accurate results than local approaches. The Fréchet distance is a global similarity measure between curves, see e.g., the seminal paper [6]. Commonly, the Fréchet distance is illustrated as follows: Suppose a person wants to walk along one curve and his/her dog on another; the person is keeping the dog at a leash. Both person and dog walk, from starting point to ending points along their respective curves. The standard Fréchet distance is the minimum leash length required without either person or dog needing to backtrack. The weak Fréchet distance is a variant of the standard Fréchet distance in which backtracking on one or both curves is allowed. Alt and Godau [6] proposed algorithms to compute the standard and weak Fréchet distances in $\mathcal{O}(n^2 \log n)$ time, where n is the maximum number of segments in the input polygonal curves. Har-Peled and Raichel [15] showed that the weak Fréchet distance can be computed in quadratic time.

In [5], Alt et al. discussed the map matching problem set in the context of the standard Fréchet distance. I.e., their algorithm finds a path in the planar graph with minimum Fréchet distance to the given trajectory. The time complexity of their algorithm is $\mathcal{O}(mn \log^2 mn)$ where m (n, respectively) is the number of edges in the input planar graph (the input polygonal curve, respectively). Brakatsoulas et al. [7], extended the map matching algorithm of [5] for the weak Fréchet distance. The time complexity of their algorithm is $\mathcal{O}(mn \log mn)$. In [8], Chen et al. proposed a $(1 + \varepsilon)$-approximation algorithm for the map matching problem when the similarity measure is the standard Fréchet distance and input model is more "realistic". They assumed that the input polygonal curve is c-packed and the input graph is ϕ-low density in $\mathbb{R}^d$ (see Sect. 2 of [8]).

In [10], Gheibi et al. studied a natural optimization problem on the weak Fréchet distance, called the *minimum backward Fréchet distance (MBFD)* problem. There, the task is to determine a pair of walks for a given input leash length such that the total length of backtracking on both input polygonal curves is minimized. The cost of backtracking could represent, for example, the cost of moving against a flow, or the cost for a moving entity (e.g., a human, a humanoid robot) to move backwards because of the entity's physiology [9]. They proposed an algorithm solving this problem within time complexity $\mathcal{O}(n^2 \log n)$ and space complexity $\mathcal{O}(n^2)$, where n is the maximum number of segments in the polygonal curves. In [11], the weighted variant of the MBFD problem is solved in $\mathcal{O}(n^3)$ time. In this variant, each edge of the input polygonal curves has an associated non-negative weight to capture different costs for backward movement.

In this paper, we study the map matching problem when the similarity measure is the MBFD. More specifically, as input, we are given: a planar graph, H, with a straight-line embedding in a plane, a directed polygonal curve, T, and a distance $\varepsilon > 0$. As motions, both forward and backward motions along T and the edges of H are allowed. The objective is to find a path, P, in H, and a parameterization of T, that minimize the sum of the walk lengths along T and P while keeping a leash length of at most ε. We restrict the start and end point of P to be at a vertex of H. However, P may partially contain an edge of H. The difference between this problem setting and that optimization setting of [5,7], is that here the total walking length along T and P is minimized while in the other settings the leash length is minimized. The optimization problem introduced in this paper, can also be used to track objects moving on road networks. To ensure high-quality tracking, the mobile tracker must remain within a distance of ε to the moving object, at all time. To minimize energy consumption, the tracker wants to minimizes walking distance. This type of scenario has been discussed in the context of wireless networks (see [12,13]).

Figure 1 shows an example of an embedding of a planar graph, H, in $\mathbb{R}^2$, a polygonal curve, T, and a length ε. The dog walks on T from $T(0)$ to $T(4)$ and the person chooses a path in H, from one vertex of H to another vertex. Two points, a and b, are determined on $\langle v_4, v_5 \rangle$ and $\langle v_3, v_4 \rangle$ respectively. A path in H, and a walk on T, that minimize the walking lengths on H and T, are as follows: the dog starts at $T(0)$ and continues on T. The person starts at v_1 and walks on the edges $\langle v_1, v_3 \rangle$, $\langle v_3, v_4 \rangle$, and $\langle v_4, v_5 \rangle$. They move together in a forward direction until the dog reaches the end of the second segment of T and the person reaches the point a on $\langle v_4, v_5 \rangle$. Then, the dog continues to move forwards until the end of the third segment of T is reached, while the person moves backwards from a to v_4 and then to b. At the final step, they move forwards again together until the dog reaches the end of T and the person reaches v_5. We show the path in the graph by the sequence of its vertices, $P^* = [v_1, v_3, v_4, a, v_4, b, v_4, v_5]$.

The structure of this paper is as follows. In Sect. 2, we discuss preliminaries and define the problem formally. In Sect. 3, we propose a polynomial time algorithm for the map matching problem introduced. Then, in Sect. 4, we develop an algorithm with improved time and space complexities. In Sect. 5, we sketch a solution to a weighted problem variant. Finally, in Sect. 6, we conclude the paper.

2   Preliminaries and Definitions

A geometric path in $\mathbb{R}^2$ is a sequence of points in 2D Euclidean space, $\mathbb{R}^2$. A polygonal curve, or a discrete geometric path, is a geometric path, sampled by a finite sequence of points (called vertices), which are connected by line segments (called edges) in order. Let $T : [0, n] \to \mathbb{R}^2$ be a polygonal curve with n segments. A vertex of T is denoted by $T(i)$, $i = 0, \ldots, n$. Let $H = \langle V_H, E_H \rangle$ be a planar graph with a straight-line embedding in $\mathbb{R}^2$ where V_H (E_H, respectively) is the set of vertices (edges, respectively) of H. In this paper, the geometric embedding of H is crucial and we simply refer to the straight-line embedding of the graph in $\mathbb{R}^2$ as H. A path, P, in H, is a polygonal curve $P : [0, 1] \to H$, such that $P \subset H$ and $P(0), P(1) \in V_H$.

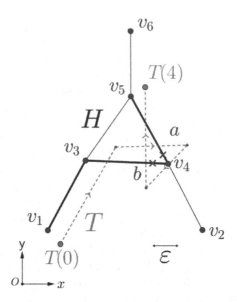

Fig. 1. An embedding of a planar graph, H, a polygonal curve, T, and a length ε are given. The path $P^* = [v_1, v_3, v_4, a, v_4, b, v_4, v_5]$, in H, is a part of a solution to the map matching problem instance. The edges of H that P^* lies on, are illustrated in bold.

A parameterization of a polygonal curve, $T : [0, n] \rightarrow \mathbb{R}^2$, is a continuous function $f : [0, 1] \rightarrow [0, n]$, where $f(0) = 0$ and $f(1) = n$. Note that a parameterization is corresponding to a walk on T and the interval $[0, 1]$ is representing time during the walk.

Walking Length. Let f be a parameterization of a polygonal curve, T. Let $\mathcal{D}_f \subseteq [0, 1]$ be the closure of the set of times in which $f(t)$ is decreasing (i.e., the movement is backward). The walking length of T is defined by Formula 1, where $\|.\|$ is the Euclidean norm and $(.)'$ is derivative.

$$\mathcal{L}_f(T) := \|T\| + 2 \int_{t \in \mathcal{D}_f} \|(T(f(t)))'\| dt \tag{1}$$

Note that if f is monotone (i.e., there is no backward movements on T), then $\mathcal{L}_f(T) = \|T\|$.

Problem Definition. Suppose H, T and a length, $\varepsilon > 0$, are given. The objective is to find a path in H and a parameterization of T such that sum of the length of P and the walking length of T is minimized (Formula 2). We consider only paths in the graph and parameterizations of T that guarantee to maintain the leash length at most ε, during the walks.

$$\mathcal{M}^\varepsilon(H, T) := \inf_{P \subset H, f} \{\|P\| + \mathcal{L}_f(T)\}. \tag{2}$$

Deformed Free-Space Surface. The free-space diagram is a structure, used to decide whether the Fréchet distance between two polygonal curves is upper bounded by a given ε [6]. In [5], Alt et al. introduced a 3D structure, called free-space surface, to solve the decision version of their map matching problem. Here, we use both free-space diagram and free-space surface. However, we modify them slightly, to fit our problem setting.

Let $P : [0,1] \to H$ be a path in H with $k+1$ vertices, $[p_0, p_1, \ldots, p_k]$. The *free-space diagram* is the rectangle $[0,1] \times [0,1]$, partitioned into n columns and k rows. It consists of nk parameter cells $C^{x,y}$, for $x = 1, \ldots, n$ and $y = 1, \ldots, k$. Cell $C^{x,y}$ is the result of the product of two sub-intervals of $[0,1]$ that are mapped to edge $\overrightarrow{T(x-1)T(x)}$ of T and edge $\overrightarrow{p_{y-1}p_y}$ of P, respectively. We call a point $(t_1, t_2) \in [0,1]^2$ white if $d(T(f(t_2)), P(t_1)) \leq \varepsilon$, where d is the Euclidean distance; otherwise, we call it black. It has been shown that the set of all white points inside a cell $C^{x,y}$ is determined by the intersection of an ellipse with $C^{x,y}$. This set is called the free-space region of that cell. The boundaries of a cell and its corresponding ellipse intersect at most eight times. These intersection points form at most four intervals of white points on the boundary of the cell (i.e., at most one interval per side of the cell). Note that two adjacent cells have the same interval on the shared side between the cells. The union of all cells' free-spaces is the free-space (or white-space) of the diagram; it is denoted by W_P. The complement of W_P is the forbidden-space (or black-space) of the diagram and is denoted by B_P. We stretch/compress the columns and rows of the free-space diagram, such that their widths and heights are equal to the lengths of the corresponding segments of T and P, respectively. The resulting diagram is called the *deformed free-space diagram* and is denoted by $\mathcal{F}_\varepsilon(T, P)$. In Fig. 2,

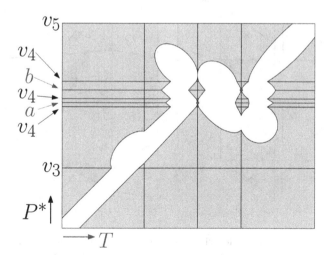

Fig. 2. The free-space diagram $\mathcal{F}_\varepsilon(T, P^*)$ is drawn. W_P is the white area and B_P is the gray area.

the free-space diagram $\mathcal{F}_\varepsilon(T, P^*)$ is drawn, where $P^* = [v_1, v_3, v_4, a, v_4, b, v_4, v_5]$ is a path in H, denoted as a sequence of its vertices.

Note that if the path P contains only a single vertex of H, $v_i \in V_H$, then $\mathcal{F}_\varepsilon(T, P)$ is a line segment and its length is equal to the Euclidean length of T. We call this 1D free-space diagram, $\mathcal{F}_i$, the *deformed free-space line* of v_i. We denote the left endpoint of $\mathcal{F}_i$ (i.e., the endpoint corresponding to $T(0)$) by s_i and the right endpoint of $\mathcal{F}_i$ (i.e., the endpoint corresponding to $T(n)$) by t_i. If P contains only an edge, $\langle v_i, v_j \rangle \in E_H$, of H, then $\mathcal{F}_\varepsilon(T, P)$ has only one row. We call this row the *deformed free-space face* of $\langle v_i, v_j \rangle$, and denote it by $\mathcal{F}_i^j$.

Note that $\mathcal{F}_i^j$ and $\mathcal{F}_j^k$ have $\mathcal{F}_j$ in common. Therefore, gluing $\mathcal{F}_i^j$ and $\mathcal{F}_j^k$ along $\mathcal{F}_j$ produces a conforming surface. Thus, we can construct the *deformed free-space surface* as follows. We first lay out the straight-line embedding of H in the xy-plane. For each edge $\langle v_i, v_j \rangle \in E_H$, we lay out $\mathcal{F}_i^j$, orthogonal to the xy-plane, along z axis, such that $\mathcal{F}_i$ ($\mathcal{F}_j$, respectively) is on top of v_i (v_j, respectively) and s_i (s_j, respectively) is in the xy-plane. Note that $\mathcal{F}_i^j$ is stretched along z axis from the plane $z = 0$ to the plane $z = \|T\|$. Suppose $Adj(v_j)$ is the set of all

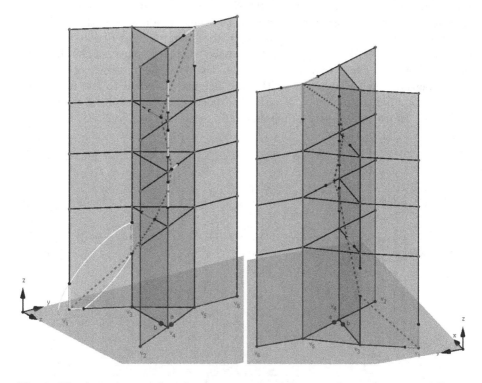

Fig. 3. The free-space surface for the example of Fig. 1 is drawn from two different viewpoints in 3D. The yellow line segments show the intervals on the cell boundaries. The red dashed polygonal curve is a path on the white-surface that realizes an optimal solution to our problem setting (Color figure online).

vertices $v_k \in V_H$ such that $\langle v_j, v_k \rangle \in E_H$. We glue $\mathcal{F}_i^j$ to $\mathcal{F}_j^k$ along $\mathcal{F}_j$, where $v_k \in Adj(v_j)$. Also, we glue $\mathcal{F}_i^j$ to $\mathcal{F}_h^i$ along $\mathcal{F}_i$, where $v_i \in Adj(v_h)$. The result is a conforming $3D$ surface between two planes, $z = 0$ and $z = \|T\|$, called deformed free-space surface and is denoted by $\mathcal{S} = H \times [0, \|T\|]$. Note that s_i is on the plane $z = 0$ and t_i is on the plane $z = \|T\|$, $i = 1, \ldots, |V_H|$. The union of the white-space (black-space, respectively) of all faces of $\mathcal{S}$ is called the *white-surface* (*black-surface*, respectively) and is denoted by $\mathcal{W}$ ($\mathcal{B}$, respectively). For the given planar graph H, the polygonal curve T, and the length ε in Fig. 1, the corresponding deformed free-space surface is shown in Fig. 3, from two points of views. Since the white-space of each cell of any $\mathcal{F}_i^j$ is convex, for simplicity, we just draw the white-space intervals on the boundary of the cells. In this figure, the red dashed polygonal curve is a path on the white-surface $\mathcal{W}$, from s_1 to t_5, that realizes $P^* = [v_1, v_3, v_4, a, v_4, b, v_4, v_5]$, in H, and a parameterization of T, that is an optimal solution to our problem setting. It intersects the following free-space faces sequentially: $\mathcal{F}_1^3$, $\mathcal{F}_3^4$, $\mathcal{F}_4^5$, $\mathcal{F}_3^4$, $\mathcal{F}_4^5$.

3   Algorithm

In this section, we first transform the map matching problem to a shortest path problem on a weighted graph, $\mathcal{G} = \langle V, E \rangle$; this yields a polynomial time algorithm. Before discussing the construction of $\mathcal{G}$, we introduce a set of Steiner points on the boundary of the cells of $\mathcal{S}$.

Steiner Points. We position Steiner points so as to create intervals on the boundary of the cells of $\mathcal{S}$. There are two types of intervals, Type 1 and Type 2. We classify the Steiner points based on the type of the intervals that they belong to. We denote the set of Type 1 (Type 2, respectively) Steiner points by S_1 (S_2, respectively).

Type 1. We say an interval is Type 1, if it lies completely in a plane, $z = c$, parallel to the xy-plane, where c is a constant. Each deformed free-space face, $\mathcal{F}_i^j$, may have $n + 1$ Type 1 intervals, $\mathcal{FI}_i^j(\ell)$, $\ell = 0, \ldots, n$, shared between its cells (where n is the number of edges in T). For each interval $\mathcal{FI}_i^j(\ell)$, we project the endpoints of $\mathcal{FI}_i^j(\ell)$ orthogonally to all $\mathcal{FI}_i^j(k)$, $k \neq \ell$. If the line segment from an endpoint of $\mathcal{FI}_i^j(\ell)$ to its projection on $\mathcal{FI}_i^j(k)$ lies in the free-space of $\mathcal{F}_i^j$ and the projection point is not identical with an endpoint of $\mathcal{FI}_i^j(k)$, then we take the projection point as a Type 1 Steiner point (see Fig. 4). The set of all Steiner points, obtained by the projections on $\mathcal{F}_i^j$, for all $\langle v_i, v_j \rangle \in E_H$, is denoted by S_1.

Type 2. We say an interval is Type 2 if it lies completely on a deformed free-space line. As we mentioned in Sect. 2, a plane $z = c$ corresponds to a point on the given trajectory T. The intersection of $z = c$ and $\mathcal{S}$ is an instance of H, denoted by H_c. Note that some part (possibly empty) of H_c is in $\mathcal{W}$. Let $z = h_j$ be the corresponding plane of $T(j)$, a vertex of T. The part of the deformed free-space surface, $\mathcal{S}$, between the two parallel planes, $z = h_{j-1}$ and $z = h_j$,

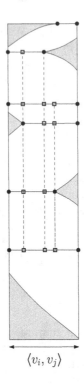

$$\langle v_i, v_j \rangle$$

Fig. 4. The free-space face $\mathcal{F}_i^j$ is drawn. The endpoints of the intervals, $\mathcal{FI}_i^j(\ell)$, are shown by points and the Type 1 Steiner points are shown by squares.

$j = 1, \ldots, n$, corresponds to edge $\overrightarrow{T(j-1)T(j)}$ of T. We denote this part of $\mathcal{S}$ by $\mathcal{T}_{j-1}^j = H \times [h_{j-1}h_j]$. In $\mathcal{T}_{j-1}^j$, $j = 1, \ldots, n$, there is at most one Type 2 interval per vertex $v_i \in V_H$. We denote these Type 2 intervals by $\mathcal{TI}_{j-1}^j(i)$, $i = 1, \ldots, |V_H|$. Suppose $z = c$ is the plane that is passing through an endpoint, p, of $\mathcal{TI}_{j-1}^j(i)$. Let the intersection of $z = c$ with $\mathcal{TI}_{j-1}^j(k)$, $k \neq i$, be q_k. Note that both p and q_k are on the graph H_c. Then, if q_k is not an endpoint of $\mathcal{TI}_{j-1}^j(k)$ and there is a path, from p to q_k, in H_c, that is in $\mathcal{W}$, then q_k is a Type 2 Steiner Point. The set of all Type 2 Steiner points is denoted by S_2. An example is given in Fig. 5. Suppose it is $\mathcal{T}_{j-1}^j$, for $j = 1$. In this example, there are four yellow intervals, $\mathcal{TI}_{j-1}^j(1)$, $\mathcal{TI}_{j-1}^j(3)$, $\mathcal{TI}_{j-1}^j(5)$, and $\mathcal{TI}_{j-1}^j(6)$. The black points show the interval endpoints and red points show the Type 2 Steiner points. For simplicity, only two, out of eight planes, are drawn. The plane z_3 (z_6, respectively) is passing through an endpoint of $\mathcal{TI}_{j-1}^j(3)$ ($\mathcal{TI}_{j-1}^j(6)$, respectively). The intersections of z_3 with $\mathcal{TI}_{j-1}^j(5)$ and $\mathcal{TI}_{j-1}^j(6)$ are Steiner points. However, the intersection of z_6 with $\mathcal{TI}_{j-1}^j(3)$ is not a Steiner point.

Constructing Graph. Now, we explain the construction of $\mathcal{G} = \langle V, E \rangle$. Recall that the white-surface (the white-space of $\mathcal{S}$) is denoted by $\mathcal{W}$. The vertices of $\mathcal{W}$ are the end points of the intervals on the boundary of the cells in $\mathcal{S}$ (at most

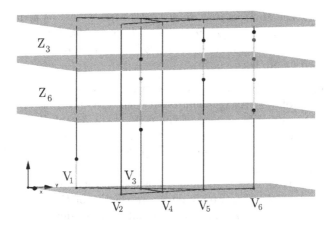

Fig. 5. An example of T^j_{j-1}, for $j = 1$, is drawn. In this example, there are four intervals that are shown by yellow color. The black balls show the interval endpoints and red balls show the Type 2 Steiner points (Color figure online).

4 intervals may exist per cell). We denote the set of vertices of $\mathcal{W}$ by $V_{\mathcal{W}}$. The set of vertices, V, of $\mathcal{G}$, is $V = V_{\mathcal{W}} \cup S_1 \cup S_2$. Note that V contains all s_i and t_i if they are in $\mathcal{W}$. Every two vertices, $v_1, v_2 \in V$, that are on the boundary of a cell, are linked by two directed edges in E, from v_1 to v_2, $\langle v_1, v_2 \rangle$, and vice versa, $\langle v_2, v_1 \rangle$. The weight of an edge $e = \langle v_1, v_2 \rangle \in E$, is its length in the L_1 metric, $|e|_1$.

Obtain an Optimal Solution. In order to have an optimal solution, at least one s_i and one $t_j, i, j = 1, \ldots, |V_H|$, must be in $\mathcal{W}$. The main steps of the algorithm are as follows:

– Find all the vertices in V_H that are in ε distance of $T(0)$ ($T(n)$, respectively), $v_{i_1}, \ldots, v_{i_{k_1}}$ ($v_{j_1}, \ldots, v_{j_{k_2}}$, respectively).
– Add an extra node, s', to $\mathcal{G}$, and add k_1 extra directed edges, $\langle s', s_{i_1} \rangle, \ldots, \langle s', s_{i_{k_1}} \rangle$, to E. The weight of these k_1 edges are set to zero. Analogously, add another extra node, t', to $\mathcal{G}$, and add k_2 extra directed edges, $\langle t_{j_1}, t' \rangle, \ldots, \langle t_{j_{k_2}}, t' \rangle$, to E. The weight of these k_2 edges are also set to zero.
– Find a shortest path, from s' to t', in $\mathcal{G}$. Note that if there is no path from s' to t' in $\mathcal{G}$, then there is no solution for the given leash length.
– Remove s' and t' from the head and tail of the shortest path. The remaining path is from one s_i to one t_j. It gives an optimal solution to our problem setting.

Note that, a vertex of $\mathcal{G}$ (except s' and t') is also represented by a point in $\mathcal{W}$. Therefore, the geometric embedding of a path, from one s_i to one t_j, in $\mathcal{G}$, is constructed by connecting the consecutive vertices of the path in $\mathcal{W}$ by line segments.

Observation 1. *Let Π be a path in the white-space, $\mathcal{W}$, of a deformed free-space surface, $\mathcal{S}$, from one s_i to one t_j. Π realizes a path, $P : [0,1] \to H$, in H, and a parameterization, $f : [0,1] \to [0,n]$, of T, that maintain the leash length at most ε, for all $t \in [0,1]$.*

Constructing a Path in H. We can construct a path P in H, from the given path Π in $\mathcal{W}$, as follows. As we mentioned earlier in this section, we have two types of intervals on the boundary of the cells, Type 1 and Type 2. A Type 1 interval lies completely on a plane, $z = c$, parallel to the xy-plane. A Type 2 interval lies completely on a deformed free-space line, $\mathcal{F}_i$. The path Π intersects a sequence of intervals (of both types). The path P in H is constructed by processing the intervals in this sequence. For each interval in this sequence, if it is Type 2 interval, on $\mathcal{F}_i$, then we append v_i to the tail of P. If it is Type 1, then the intersection point, q, of Π and that interval, is appended to the tail of P, as a vertex of P. Note that q may not be a vertex of H. However, it is a point on an edge of H. At the end, we connect the consecutive vertices in P by straight line segments.

Correctness. To establish the correctness, we use norms in two spaces: (1) the Euclidean space of the embedding of the input graph and the polygonal curve, called the input space, (2) the deformed free-space surface, called the configuration space. In the input space, we denote the Euclidean length of a polygonal curve T by $\|T\|$. We also defined walking length of T, $\mathcal{L}_f(T)$, based on a parameterization f. Note that if f is a monotone parameterization, then $\mathcal{L}_f(T) = \|T\|$. In configuration space, a path from an s_i to a t_j in $\mathcal{W}$, is also denoted by its vertices, $\Pi : \langle s_i = p_1, p_2, \ldots, p_k = t_j \rangle$. The length, $|.|_1$, of each segment of Π is calculated by the L_1 metric. The length of a path, $|\Pi|_1$, is the sum of the length of its segments.

Lemma 1 is at the heart of the correctness proof. This section is concluded by a corollary to Lemma 1 and Observation 1, that is, in order to find a solution for our problem setting, it suffices to find a shortest path from s' to t' in $\mathcal{G}$.

Lemma 1. *For any path $\Pi : \langle s_i = p_1, p_2, \ldots, p_{k_1} = t_j \rangle$ in $\mathcal{W}$, there is a path $\Pi' : \langle s_i = p'_1, p'_2, \ldots, p'_{k_2} = t_j \rangle$ in $\mathcal{W}$ such that $\Pi' \subset \mathcal{G}$ and $|\Pi'|_1 \leq |\Pi|_1$.*

Proof. The path Π intersects a sequence, SF, of deformed free-space faces. Every two consecutive faces in SF share a deformed free-space line. Therefore, we can unfold the free-space faces in the sequence, along the shared free-space lines. The result is a $2D$ free-space diagram, denoted by $\mathcal{F}_\varepsilon(SF)$. W.l.o.g., we can assume that $\mathcal{F}_\varepsilon(SF)$ is axis aligned in $\mathbb{R}^2$. The path Π is also unfolded into a $2D$ path in the white-space, $\mathcal{W}_{SF}$, of $\mathcal{F}_\varepsilon(SF)$. Note that unfolding does not change the length of a path. As an example, in Fig. 6a, the result of unfolding the faces that are intersected by the red dashed polygonal curve in Fig. 3, is shown.

Let $\Pi_{opt} : \langle s_i = q_1, q_2, \ldots, q_{k_3} = t_j \rangle$ be a L_1 shortest path, from s_i to t_j, in $\mathcal{W}_{SF}$. Then, $|\Pi_{opt}|_1 \leq |\Pi|_1$. To prove the lemma, it suffices to show that there is a path $\Pi' \subset \mathcal{G} = \langle V, E \rangle$, from s_i to t_j, in $\mathcal{W}_{SF}$, such that $|\Pi'|_1 = |\Pi_{opt}|_1$.

We know that the vertices of Π_{opt} are endpoints of some intervals on the boundary of the cells of $\mathcal{F}_\varepsilon(SF)$ [14] (the well known rubber band property of

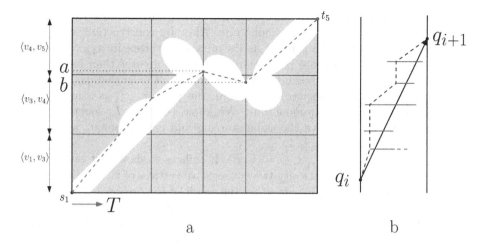

Fig. 6. (a) The result of unfolding the sequence of deformed free-space faces that are intersected by the red dashed polygonal curve in Fig. 3. It is a $2D$ free-space diagram, $\mathcal{F}_\varepsilon(SF)$. The red dashed polygonal curve is shown after unfolding. (b) Illustration of case 1 in the proof of Lemma 1 (Color figure online).

shortest paths). Therefore, the vertices of Π_{opt} are in V (i.e., the set of vertices of $\mathcal{G}$). Thus, it is sufficient to show that for each edge, $\overrightarrow{q_i q_{i+1}}$, of Π_{opt}, there is a path, $\pi_{q_i q_{i+1}}$, from q_i to q_{i+1}, in $\mathcal{G}$, that $|\overrightarrow{q_i q_{i+1}}|_1 = |\pi_{q_i q_{i+1}}|_1$. Two cases arise depending on whether $\overrightarrow{q_i q_{i+1}}$ lies completely within a row (or a column) of $\mathcal{F}_\varepsilon(SF)$, or not. We need a definition before discussing these cases. We assume that $\mathcal{F}_\varepsilon(SF)$ is an axis-aligned rectangle in a $2D$ Cartesian coordinate system, where the x-axis corresponds to T (Fig. 6a). We say a path $\Pi \in \mathcal{W}_{SF}$ is x-monotone (y-monotone, respectively), if any vertical (horizontal, respectively) line intersects it at most ones. Π is said to be xy-monotone, if it is both x- and y-monotone.

Case 1. In this case, $\overrightarrow{q_i q_{i+1}}$ lies completely within a column (or a row) of $\mathcal{F}_\varepsilon(SF)$. Here, we discuss the case when it lies within a column (see Fig. 6b); the arguments are analogous for case of a row. W.l.o.g we assume that $\overrightarrow{q_i q_{i+1}}$ is xy-increasing. The other cases are symmetric. Edge $\overrightarrow{q_i q_{i+1}}$ intersects a sequence of horizontal intervals, I_z, within a column. They are sorted based on their y coordinates. We construct $\pi_{q_i q_{i+1}}$ sequentially and always denote the last vertex appended to $\pi_{q_i q_{i+1}}$ by π_{last}. Initially, $\pi_{q_i q_{i+1}}$ contains only q_i and $\pi_{last} = q_i$. The sequence of intervals are processed sequentially. Suppose we processed interval I_z and now we want to process I_{z+1}. We project orthogonally from π_{last} to I_{z+1}. If the projection point exists (i.e., the perpendicular line from π_{last} to I_{z+1} intersects I_{z+1}), then append the projection point on I_{z+1} to $\pi_{q_i q_{i+1}}$ and update π_{last}. Otherwise, the closest endpoint of I_{z+1} to π_{last} is appended to $\pi_{q_i q_{i+1}}$ and we update π_{last}. When all intervals, I_z, have been processed, q_{i+1} is appended to $\pi_{q_i q_{i+1}}$.

Since the sorted list of intervals withing a column are traversed by $\pi_{q_i q_{i+1}}$ sequentially, the path $\pi_{q_i q_{i+1}}$ is y-monotone. Also, by construction, each vertex of $\pi_{q_i q_{i+1}}$ either has the same x as its preceding vertex in $\pi_{q_i q_{i+1}}$ (i.e., it is the result of the orthogonal projection) or its x is greater than its preceding vertex's x (since the orthogonal projection does not exist and $\overrightarrow{q_i q_{i+1}}$ is xy-increasing inside the white-space). Therefore, the path $\pi_{q_i q_{i+1}}$ is x-monotone. Thus, the path $\pi_{q_i q_{i+1}}$ is xy-monotone. We know that the L_1 length of two xy-monotone paths that have the same starting and ending points, are equal. Therefore, $|\overrightarrow{q_i q_{i+1}}|_1 = |\pi_{q_i q_{i+1}}|_1$.

Now, we prove that $\pi_{q_i q_{i+1}} \subset \mathcal{G} = \langle V, E \rangle$. It suffices to show that each vertex of $\pi_{q_i q_{i+1}}$ is in V and between every two consecutive vertices of $\pi_{q_i q_{i+1}}$ there is an edge in E. Each vertex of $\pi_{q_i q_{i+1}}$ is either the result of the orthogonal projection or an endpoint of an interval. Therefore, each vertex is either a Steiner point or a vertex of the white-surface. In both cases, the vertex is in V. In addition, between every two consecutive vertices of $\pi_{q_i q_{i+1}}$ there is an edge in E because every two consecutive vertices of $\pi_{q_i q_{i+1}}$ lie on the boundary of a cell and, by the construction of $\mathcal{G}$, all members of V that lie on the boundary of a cell are linked by edges in E.

Case 2. In [11], Sect. 4, Lemma 4, it is proved that if $\overrightarrow{q_i q_{i+1}}$ does not lie completely within a row and within a column of $\mathcal{F}_\varepsilon(SF)$, then there is a xy-monotone path $\pi'_{q_i q_{i+1}}$, from q_i to q_{i+1}, such that its edges lie completely within a row and within a column of $\mathcal{F}_\varepsilon(SF)$. For each edge of $\pi'_{q_i q_{i+1}}$, we apply case 1. Then, we concatenate the resulting xy-monotone paths for edges of $\pi'_{q_i q_{i+1}}$, to obtain $\pi_{q_i q_{i+1}}$. Since, xy-monotone paths for edges of $\pi'_{q_i q_{i+1}}$ are in $\mathcal{G}$ (as we proved in Case 1), the resulting path, $\pi_{q_i q_{i+1}}$, is a xy-monotone path in $\mathcal{G}$. Therefore, $|\overrightarrow{q_i q_{i+1}}|_1 = |\pi_{q_i q_{i+1}}|_1$.

Corollary 1. *For any pair of s_i and t_j, if t_j is reachable from s_i by a path in $\mathcal{W}$, then there is a path from s_i to t_j, in $\mathcal{G}$, that is a L_1 shortest path in $\mathcal{W}$.*

Corollary 2. *A shortest path in $\mathcal{G}$, from s' to t', yields an optimal solution for our problem setting.*

Proof. Let Π'_{opt} be a shortest path in $\mathcal{G}$, from s' to t'. We remove s' and t' from the head and tail of Π'_{opt}. The result, Π_{ij}, is a shortest path from s_i to t_j. Therefore, among all possible shortest paths $\Pi_{k\ell}$, for s_k and t_ℓ, $k, \ell = 1, \dots, |V_H|$, the pair (s_i, t_j) has a shortest L_1 shortest path, Π_{ij}. By Corollary 1, $\Pi_{ij} \subset \mathcal{G}$ is a L_1 shortest path in $\mathcal{W}$. Each point on Π_{ij} is corresponding to a point, p, on H and a point, q, on T, such that the Euclidean distance of p and q is less than ε. By Observation 1, Π_{ij}, is corresponding to a path, P, in H, from $v_i \in V_H$ to $v_j \in V_H$, and a parameterization, f, of T. The summation of the Euclidean length of P, $\|P\|$, and the walking length of T, $\mathcal{L}_f(T)$, is equal to the L_1 length of Π_{ij}. Since Π_{ij} is a shortest L_1 shortest path, P and f minimize the matching cost, $\mathcal{M}^\varepsilon(H, T)$ Eq. (2).

Theorem 1. *Let H be a planar graph with a straight-line embedding in a plane, T be a directed polygonal curve, and $\varepsilon > 0$ be a distance. A path, $P : [0,1] \to H$,* *between two vertices of H, and a parameterization, f, of T, that minimize the sum of the walking length of T and P, can be found in polynomial time and space. It is guaranteed that at any time $t \in [0,1]$, the Euclidean distance between $P(t)$ and $T(f(t))$ is at most ε.*

Proof. The correctness follows directly from Corollary 2. The deformed free-space surface, $\mathcal{S}$, has $\mathcal{O}(mn)$ cells, where m (n, respectively) is the number of edges of H (T, respectively). Each cell of $\mathcal{S}$ has at most four intervals and at most $\mathcal{O}(m+n)$ Steiner points on its intervals. Therefore, the graph $\mathcal{G}$ has $\mathcal{O}\left(mn(m+n)\right)$ vertices and $\mathcal{O}\left(mn(m+n)^2\right)$ edges (including the extra edges that connect s' and t' to the graph). In addition, it takes $\mathcal{O}\left(n^2\right)$ time to compute all Type 1 Steiner points for each free-space face. Therefore, computing S_1 takes $\mathcal{O}\left(mn^2\right)$ time. In order to compute Type 2 Steiner points, we use breadth first search for each interval endpoint to propagate the projection on the instance of the graph, H_c, in the plane $z = c$. Therefore, computing S_2 takes $\mathcal{O}\left(nm^2\right)$ time. At the end, it is possible to find a shortest path in $\mathcal{G}$, from s' to t', in $\mathcal{O}\left(mn(m+n)^2\right)$ time, by using Dijkstra's algorithm. Therefore, both the total time and space complexities are $\mathcal{O}\left(mn(m+n)^2\right)$.

4   Improvement

In Sect. 3, we showed that the graph $\mathcal{G} = \langle V, E \rangle$ contains a path that yields an optimal solution for our problem setting. The bottleneck in the time complexity of the algorithm in Sect. 3 is due to the number of edges of $\mathcal{G}$. In this section, we construct a new graph $\mathcal{G}' = \langle V, E' \rangle$, such that $|E'| < |E|$ and it preserves the connectivity information of $\mathcal{G}$. More precisely, if there is a path, from $v_i \in V$ to $v_j \in V$, in $\mathcal{G}$, then there is a path, from v_i to v_j, in $\mathcal{G}'$, with the same L_1 length.

Based on the construction of $\mathcal{G}$, there are at most $\mathcal{O}(m+n)$ vertices in V (including the interval endpoints and Steiner points) on the boundary of each cell, C, of $\mathcal{S}$. We connect these $\mathcal{O}(m+n)$ vertices by a linear number of edges, in E', as follows. The weight of each edge in E' is equal to its L_1-length. Let T, B, L, and R be the intervals on the top, bottom, left and right side of C, respectively. Suppose cell C is in a $2D$ Cartesian coordinate system and the vertices on each interval $I \in \{T, B, L, R\}$ are sorted by x and y. Every two adjacent vertices, v_i and v_{i+1}, on I, are linked by two directed edges, $\langle v_i, v_{i+1} \rangle$ and $\langle v_{i+1}, v_i \rangle$ (Fig. 7). Every two of the eight interval endpoints are linked by two directed edges assuming they are not identical. A vertex v_i on interval L (T, respectively), is linked by two directed edges to another vertex v'_i on R (B, respectively) if v'_i has the same y (x, respectively) coordinate as v_i; the two edges are denoted by $\langle v_i, v'_i \rangle$ and $\langle v'_i, v_i \rangle$, respectively. By this approach, each vertex of $\mathcal{G}'$ on the boundary of C is connected to a constant number of vertices of $\mathcal{G}'$ on the boundary of C. It is now straightforward to prove the following lemma.

Lemma 2. *Let v_i and v_j be two vertices, in V, on the boundary of a cell, C, of $\mathcal{S}$. There is a path, from v_i to v_j, in $\mathcal{G}'$, that has the same L_1 length as the direct line segment between them.*

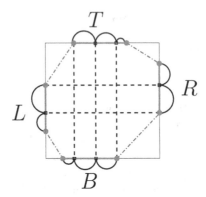

Fig. 7. A cell of the free-space surface is drawn. The red solid line segments show the four intervals on the boundary of the cell. The arcs show the edges in E0 that connect every two adjacent vertices of G0, on each interval. The dashed black line segments show the edges in E0 that connect a vertex with its orthogonal projection on the opposite side of the cell. The dash dotted blue line segments show some of the edges that connect endpoints of the intervals. For simplicity, we did not draw all of them (Color figure online).

Corollary 3. *There is a path in $\mathcal{G}'$ that realizes an optimal solution for our problem setting.*

Theorem 2. *Let H be a planar graph with a straight-line embedding in a plane, T be a directed polygonal curve, and $\varepsilon > 0$ be a distance. A path, $P : [0, 1] \to H$, between two vertices of H, and a parameterization, f, of T, that minimize the sum of the walking length of T and P, can be found in $\mathcal{O}(nm(n + m) \log(nm))$ time and $\mathcal{O}(nm(n + m))$ space, where n (m, respectively) is the number of edges of T (H, respectively). It is guaranteed that at any time $t \in [0, 1]$, the Euclidean distance between $P(t)$ and $T(f(t))$ is at most ε.*

Proof. The correctness follows directly from Corollary 3. The number of vertices and edges of $\mathcal{G}'$ (and the total space complexity) is upper-bounded by $\mathcal{O}(nm(n + m))$. Using Dijkstra's algorithm, we find a shortest path in $\mathcal{G}'$, from s' to t'. Therefore, the time complexity of our algorithm is $\mathcal{O}(nm(n + m) \log(nm))$. Note that if there is no pair of (s_k, t_ℓ), $k, \ell = 1, \ldots, |V_H|$, in a connected component of $\mathcal{G}'$, then there is no feasible solution.

5   Weighted Non-planar Graphs

We assumed that the input graph H is planar. That makes the illustration of the algorithm easier since the faces of the free-space surface do not intersect except at the boundary of the faces. However, all lemmas and theorems, derived in Sects. 3 and 4, are proved without making an assumption regarding the planarity of H. Therefore, the algorithm proposed in this paper remains correct for any graph for which a straight-line embedding in a plane is provided (see [5], Sect. 2.7). In the embedding, the edges of the graph may intersect. Transition from one edge to another is allowed only at a vertex.

We also assumed that H is unweighted. Here, we sketch how the proposed algorithm can be generalized to also handle the problem instance, when H is weighted. Suppose that each edge of H has a non-negative, real weight. A weight could represent the cost of moving on the edge of the graph. The edges of the input polygonal curve T could also have weights capturing the costs of moving forwards and backwards. The objective is to find a path in H whose weighted walking length is minimized. In the weighted problem setting, inside each cell of the free-space surface, there are two weights, one corresponding to an edge of T and one corresponding to an edge of H. These weights are fixed inside the cell and do not change. Therefore, in the construction of $\mathcal{G}$ or $\mathcal{G}'$, instead of computing the L_1 length for each edge, e, we compute the orthogonal projections of e onto H and T. Then, we multiply the projection lengths with the corresponding weights, and the sum of these multiplications is the weight that we assign to e. The remaining parts of the algorithm remains the same and the time and space complexities do not change.

6   Conclusion

In this paper, we discussed a geometric algorithm for the map matching problem that minimizes the walking length. We established that this problem setting is dual to a weighted shortest path problem. Then, we proposed an algorithm with $\mathcal{O}\left(mn\left(m+n\right)\log(mn)\right)$ time and $\mathcal{O}\left(mn\left(m+n\right)\right)$ space complexities, where m (n, respectively) is the number of edges of H (T, respectively). At the end, we discussed that the proposed algorithm is easily adaptable to handle weighted non-planar graphs. It is still open if we can improve the proposed algorithm further, for planar graphs. The main challenge here is the existence of cycles in the input graph and propagation through the cycles.

Acknowledgment. The authors would like to thank Carola Wenk for suggesting this topic and constructive comments, and Omid Gheibi for valuable discussions.

References

1. Zheng, Y.: Trajectory data mining: an overview. ACM Trans. Intell. Syst. Technol. **6**(3), 29, Article 1 (2015)
2. Chen, B., Yuan, H., Li, Q., Lam, W., Shaw, S., Yan, K.: Map-matching algorithm for large-scale low-frequency floating car data. Int. J. Geogr. Inf. Sci. **28**(1), 22–38 (2014)
3. Ruan, F., Deng, Z., An, Q., Wang, K., Li, X.: A method of map matching in indoor positioning. In: Sun, J., Jiao, W., Wu, H., Lu, M. (eds.) CSNC 2014 Proceedings: Volume III. Lecture Notes in Electrical Engineering, vol. 305, pp. 669–679. Springer, Berlin (2014)

4. Asakura, K., Takeuchi, M., Watanabe, T.: A pedestrian-oriented map matching algorithm for map information sharing systems in disaster areas. Int. J. Know. Web Intel. **3**(4), 328–342 (2012)
5. Alt, H., Efrat, A., Rote, G., Wenk, C.: Matching planar maps. In: Proceeding of the Fourteenth Annual ACM-SIAM Symposium on Discrete Algorithms, pp. 589–598 (2003)
6. Alt, H., Godau, M.: Computing the Fréchet distance between two polygonal curves. Int. J. Comput. Geom. Appl. **5**, 75–91 (1995)
7. Brakatsoulas, S., Pfoser, D., Salas, R., Wenk, C.: On map-matching vehicle tracking data. In: Proceeding of VLDB, pp. 853–864. ACM (2005)
8. Chen, D., Driemel, A., Guibas, L., Nguyen, A., Wenk, C.: Approximate map matching with respect to the Fréchet distance. In: Proceeding of 13th ALENEX, pp. 75–83 (2011)
9. Flynn, T., Connery, S., Smutok, M., Zeballos, R., Weisman, I.: Comparison of cardiopulmonary responses to forward and backward walking and running. Med. Sci. Sports Exerc. **26**(1), 89–94 (1994)
10. Gheibi, A., Maheshwari, A., Sack, J.-R., Scheffer, C.: Minimum backward Fréchet distance. In: Proceedings of the 22nd ACM SIGSPATIAL, pp. 381–388 (2014)
11. Gheibi, A., Maheshwari, A., Sack, J.-R.: Weighted minimum backward Fréchet distance. Accepted to 27th CCCG, Kingston (2015)
12. Bhuiyan, M.Z.A., Wang, G., Vasilakos, A.V.: Local area prediction-based mobile target tracking in wireless sensor networks. IEEE Trans. Comput. **64**(7), 1968–1982 (2015)
13. Vachhani, H.: Continuous spatio temporal tracking of mobile targets, Master's thesis, Arizona State University (2014)
14. Ghosh, S.K., Mount, D.M.: An output-sensitive algorithm for computing visibility graphs. SIAM J. Comput. **20**(5), 888–910 (1991)
15. Har-Peled, S., Raichel, B.: The Fréchet distance revisited and extended. In: Proceedings of the 27th ACM SoCG, pp. 448–457 (2011)

Rainbow Domination and Related Problems on Some Classes of Perfect Graphs

Wing-Kai Hon[1]([✉]), Ton Kloks[1], Hsiang-Hsuan Liu[1,2], and Hung-Lung Wang[3]

[1] Department of Computer Science, National Tsing Hua University, Hsinchu, Taiwan
{wkhon,hhliu}@cs.nthu.edu.tw
[2] University of Liverpool, Liverpool, England
hhliu@liv.ac.uk
[3] Institute of Information and Decision Sciences,
National Taipei University of Business, Taipei, Taiwan
hlwang@ntub.edu.tw

Abstract. Let $k \in \mathbb{N}$ and let G be a graph. A function $f : V(G) \to 2^{[k]}$ is a rainbow function if, for every vertex x with $f(x) = \varnothing$, $f(N(x)) = [k]$, where $[k]$ denotes the integers ranging from 1 to k. The rainbow domination number $\gamma_{kr}(G)$ is the minimum of $\sum_{x \in V(G)} |f(x)|$ over all rainbow functions. We investigate the rainbow domination problem for some classes of perfect graphs.

1 Introduction

In 2008, Brešar et al. [2] introduced the k-rainbow domination problem, which is a generalized formulation of graph domination. In the graph domination problem, a set of vertices is selected as the "guards" such that each vertex not selected has a guard as a neighbor; while in the k-rainbow domination problem, k different types of guards are required in the neighborhood of a non-selected vertex. The k-rainbow domination-problem drew our attention because it is solvable in polynomial time for classes of graphs of bounded rankwidth but, unless one fixes k as a constant, it seems not formulatable in monadic second-order logic. Let us start with the definition.

Definition 1. *Let $k \in \mathbb{N}$ and let G be a graph. A function $f : V(G) \to 2^{[k]}$ is a k-rainbow function if, for every $x \in V(G)$,*

$$f(x) = \varnothing \quad implies \quad \cup_{y \in N(x)} f(y) = [k].$$

The k-rainbow domination number of G is

$$\gamma_{rk}(G) = \min \{ \, \|f\| \mid f \text{ is a } k\text{-rainbow function for } G \, \},$$

$$where \quad \|f\| = \sum_{x \in V(G)} |f(x)|.$$

© IFIP International Federation for Information Processing 2016
Published by Springer International Publishing Switzerland 2016. All Rights Reserved.
M.T. Hajiaghayi and M.R. Mousavi (Eds.): TTCS 2015, LNCS 9541, pp. 121–134, 2016.
DOI: 10.1007/978-3-319-28678-5_9

We call $\|f\|$ the <u>cost</u> of f over the graph G. When there is danger of confusion, we write $\|f\|_G$ instead of $\|f\|$. We call the elements of $[k]$ the <u>colors</u> of the rainbow and, for a vertex x we call $f(x)$ the <u>label</u> of x. For a set S of vertices we write

$$f(S) = \cup_{x \in S}\, f(x).$$

It is a common phenomenon that the introduction of a new domination variant is followed chop-chop by an explosion of research results and their write-ups. One reason for the popularity of domination problems is the wide range of applicability and directions of possible research. We leave our bibliography of recent publications on this specific domination variant to the full version of this paper [20]. We refer to [28] for the description of an application of rainbow domination.

To begin with, Brešar et al. showed that, for any graph G,

$$\gamma_{rk}(G) = \gamma(G \Box K_k), \tag{1}$$

where γ denotes the domination number and where $\Box$ denotes the Cartesian product. This observation, together with Vizing's conjecture, stimulated the search for graphs for which $\gamma = \gamma_{r2}$ (see also [1,21]). Notice that, by Eq. (1) and Vizing's upperbound $\gamma_{rk}(G) \le k \cdot \gamma(G)$ [29].

Chang et al. [7] were quick on the uptake and showed that, for $k \in \mathbb{N}$, the k-rainbow domination problem is NP-complete, even when restricted to chordal graphs or bipartite graphs. The same paper shows that there is a linear-time algorithm to determine the parameter on trees. A similar algorithm for trees appears in [30] and this paper also shows that the problem remains NP-complete on planar graphs.

Notice that Eq. (1) shows that $\gamma_{rk}(G)$ is a non-decreasing function in k. Chang et al. show that, for all graphs G with n vertices and all $k \in \mathbb{N}$,

$$\min\{\, k,\ n\,\} \le \gamma_{rk}(G) \le n \quad \text{and} \quad \gamma_{rn}(G) = n.$$

For trees T, Chang et al. [7] give sharp bounds for the smallest k satisfying $\gamma_{rk}(T) = |V(T)|$.

Many other papers establish bounds and relations, e.g., between the 2-rainbow domination number and the total domination number or the (weak) roman domination number, or study edge- or vertex critical graphs with respect to rainbow domination, or obtain results for special graphs such as paths, cycles, graphs with given radius, and the generalized Petersen graphs. A detailed survey can be found in [20].

Pai and Chiu [23] develop an exact algorithm and a heuristic for 3-rainbow domination. They present the results of some experiments. Let us mention that the k-rainbow domination number may be computed, via Eq. (1), by an exact, exponential algorithm that computes the domination number. For example, this shows that the k-rainbow domination number can be computed in $O(1.4969^{nk})$ [24,25].

Whenever domination problems are under investigation, the class of strongly chordal graphs are of interest from a computational point of view. Farber showed that a minimum weight dominating set can be computed in polynomial time on strongly chordal graphs [14]. Recently, Chang et al. showed that the k-rainbow dominating number is equal to the so-called weak $\{k\}$-domination number for strongly chordal graphs [2,3,8]. A weak $\{k\}$-dominating function is a function $g : V(G) \rightarrow \{0, \ldots, k\}$ such that, for every vertex x,

$$g(x) = 0 \quad implies \quad \sum_{y \in N(x)} g(y) \geq k. \tag{2}$$

The weak domination number $\gamma_{wk}(G)$ minimizes $\sum_{x \in V(G)} g(x)$, over all weak $\{k\}$-dominating functions g. In their paper, Chang et al. show that the k-rainbow domination number is polynomial for block graphs. As far as we know, the k-rainbow domination number is open for strongly chordal graphs.

It is easy to see that, for fixed k, the k-rainbow domination problem can be formulated in monadic second-order logic. For example, a function $f : V(G) \rightarrow 2^{[k]}$ can be defined using k vertex subsets $V_1, \ldots, V_k$, such that $f(x) = \{i \mid x \in V_i\}$, and the property of k-rainbow can be formulated as

$$\exists_{V_0 \subseteq V(G)} \exists_{V_1 \subseteq V(G)} \cdots \exists_{V_k \subseteq V(G)} \quad \big[(\forall_{x \in V(G)} \exists_{i \in [k] \cup \{0\}}\, x \in V_i) \wedge$$
$$(\forall_{i \in [k]} V_0 \cap V_i = \varnothing) \wedge (\forall_{x \in V_0} \forall_{i \in [k]} N(x) \cap V_i \neq \varnothing) \big].$$

The k-rainbow domination number is the minimal value of $\sum_{i=1}^{k} |V_i|$, where $(V_1, \ldots, V_k)$ defines a k-rainbow function. This shows that, when k is fixed, the parameter is computable in linear time for graphs of bounded treewidth or rankwidth [12].

Theorem 1. *Let $k \in \mathbb{N}$. There exists a linear-time algorithm that computes $\gamma_{rk}(G)$ for graphs of bounded rankwidth.*

For example, Theorem 1 implies that, for each k, $\gamma_{rk}(G)$ is computable in polynomial time for distance-hereditary graphs, i.e., the graphs of rankwidth 1. Also, graphs of bounded outerplanarity have bounded treewidth, which implies bounded rankwidth.

A direct application of the monadic second-order theory involves a constant which is an exponential function of k. In the following section we show that, in some cases, this exponential factor can be avoided. Moreover, besides weak $\{k\}$-domination, another variant, *weak $\{k\}$-L-domination*, is also conducted. A formal definition is given later in Sect. 3. The corresponding parameter, the *weak $\{k\}$-L-domination number*, is denoted by $\gamma_{wkL}(G)$. We note here that this variant was formulated in order to solve the k-rainbow domination problem [7,8]. In more detail, the results presented in this paper (see Fig. 1) consist of

- linear-time algorithms for cographs on γ_{rk} and γ_{wk}, respecitvely (Sect. 2).
- a linear-time algorithm for trivially perfect graphs on γ_{wkL} (Sect. 3).

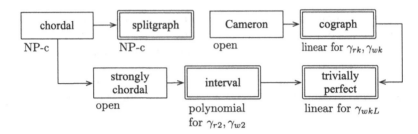

Fig. 1. Results on k-rainbow domination over graph classes. Double-lined rectangles indicate the graph classes conducted in this paper. The arrows "$\rightarrow$" denote the relation of containment, i.e., $\mathcal{A} \rightarrow \mathcal{B}$ means $\mathcal{A}$ is a superclass of $\mathcal{B}$. A graph is Cameron if it is (Bull, Gem, Co-Gem, C_5)-free [6].

– a polynomial-time algorithm for interval graphs on γ_{r2}, which is equal to γ_{w2} (Sect. 4).
– The NP-completeness of computing γ_{rk} and γ_{wk} for splitgraphs (Sect. 5).

Because of the space limitation, some proofs and related results are omitted. The details are available in the full version [20].

2 k-Rainbow Domination on Cographs

Cographs are the graphs without an induced P_4. As a consequence, cographs are completely decomposable by series and parallel operations, that is, joins and unions [15]. In other words, a graph is a cograph if and only if every nontrivial, induced subgraph is disconnected or its complement is disconnected. Cographs have a rooted, binary decomposition tree, called a cotree, with internal nodes labeled as joins and unions [11].

For a graph G and $k \in \mathbb{N}$, let $F(G, k)$ denote the set of k-rainbow functions on G. Furthermore, define

$$F^+(G, k) = \{ \, f \in F(G, k) \mid \forall_{x \in V(G)} \, f(x) \neq \varnothing \, \}$$
$$\text{and} \quad F^-(G, k) = F(G, k) \setminus F^+(G, k).$$

Theorem 2. *There exists a linear-time algorithm to compute the k-rainbow domination number $\gamma_{rk}(G)$ for cographs G and $k \in \mathbb{N}$.*

Proof. We describe a dynamic programming algorithm to compute the k-rainbow domination number. A minimizing k-rainbow function can be obtained by backtracking.

Clearly, for a k-rainbow function that has no empty-set label, the minimal cost is the number of vertices. We therefore concentrate on those k-rainbow functions for which some labels are the empty set.

For a cograph H define

$$R^+(H) = \min \{ \, \|f\|_H \ \mid \ f \in F^+(H, k) \text{ and } f(V(H)) = [k] \, \},$$
$$R^-(H) = \min \{ \, \|f\|_H \ \mid \ f \in F^-(H, k) \, \}.$$

Here, we adopt the convention that $R^-(H) = \infty$ if $F^-(H, k) = \varnothing$. Notice that

$$R^+(H) = \max \{ \, |V(H)|, \ k \, \}. \tag{3}$$

Assume that H is the union of two smaller cographs H_1 and H_2. Then clearly,

$$R^-(H) = \min \{ R^-(H_1) + |V(H_2)|, \ R^-(H_2) + |V(H_1)|, \ R^-(H_1) + R^-(H_2) \}. \tag{4}$$

Now assume that H is the join of two smaller cographs, H_1 and H_2. We claim

$$R^-(H) = \min \{ \ R^+(H_1), \ R^+(H_2), \ R^-(H_1), \ R^-(H_2), \ 2k \ \}. \tag{5}$$

To show that Eq. (5) holds, let f be a k-rainbow function from $F^-(H, k)$ with minimum cost over H. First, one can observe that

$$\|f\|_H \leq 2k \tag{6}$$

since in each of H_1 and H_2, labeling exactly one vertex with $[k]$ and the others with $\varnothing$ results in a k-rainbow function of cost $2k$.

If there exists i in $\{1, 2\}$ such that $f(x) \neq \varnothing$ for all $x \in V(H_i)$, then there is a vertex labeled with $\varnothing$ in H_{3-i}. Let $L = \cup_{y \in V(H_{3-i})} f(y)$. Define another k-rainbow function f' of H as follows. Choosing an arbitrary vertex x in $V(H_i)$, let $f'(x) = f(x) \cup L$. For $y \in V(H_i) \setminus \{x\}$, let $f'(y) = f(y)$, and for $z \in V(H_{3-i})$, let $f'(z) = \varnothing$. Notice that $f'(V(H_i)) = [k]$, and thus, f' is a k-rainbow function with cost at most $\|f\|_H$. This shows that $R^-(H) = R^+(H_i)$.

If each of H_1 and H_2 contains a vertex with label $\varnothing$, let

$$L_1 = f(V(H_1)) \quad \text{and} \quad L_2 = f(V(H_2)).$$

For each color $\ell \in [k]$, let ν_ℓ be the number of times that ℓ is used in a label, that is,

$$\nu_\ell = |\{ \, x \mid x \in V(H) \quad \text{and} \quad \ell \in f(x) \, \}|.$$

If for all ℓ, $\nu_\ell \geq 2$. It follows that $\|f\|_H \geq 2k$. Together with inequality Eq. (6), we have $R^-(H) = 2k$. Otherwise, there exists some ℓ with $\nu_\ell = 1$. Let u be the unique vertex with $\ell \in f(u)$. Assume that $u \in V(H_1)$. The case where $u \in V(H_2)$ is similar. Clearly, u is adjacent to all $x \in V(H)$ with $f(x) = \varnothing$. Modify f to f' so that $f'(u) = f(u) \cup (L_2 \setminus L_1)$, $f'(x) = f(x)$ for all $x \in V(H_1) \setminus \{u\}$, and $f'(y) = \varnothing$ for all $y \in V(H_2)$. It is not difficult to verify that f' is a k-rainbow function from $F^-(H, k)$ with cost at most $\|f\|$. Moreover, f' restricted to H_1 is a k-rainbow function with minimum cost over H_1. Thus, in this case, $R^-(H) = R^-(H_1)$. At the root of the cotree, we obtain $\gamma_{rk}(G)$ via

$$\gamma_{rk}(G) = \min \{ \ |V(G)|, \ R^-(G) \ \}.$$

The cotree can be obtained in linear time (see, e.g., [5,10,18]). Each $R^+(H)$ is obtained in $O(1)$ time via Eq. (3), and $R^-(H)$ is obtained in $O(1)$ time via Eqs. (4) and (5).
This proves the theorem. ∎

The weak $\{k\}$-domination number (recall the definition near Eq. (2)) was introduced by Brešar, Henning and Rall in [3] as an accessible, 'monochromatic version' of k-rainbow domination. In the following theorem we turn the tables.

In general, for graphs G one has that $\gamma_{wk}(G) \leq \gamma_{rk}(G)$ since, given a k-rainbow function f one obtains a weak $\{k\}$-dominating function g by defining, for $x \in V(G)$, $g(x) = |f(x)|$. The parameters γ_{wk} and γ_{rk} do not always coincide. For example $\gamma_{w2}(C_6) = 3$ and $\gamma_{r2}(C_6) = 4$. Brešar et al. ask, in their Question 3, for which graphs the equality $\gamma_{w2}(G) = \gamma_{r2}(G)$ holds. As far as we know this problem is still open. Chang et al. showed that weak $\{k\}$-domination and k-rainbow domination are equivalent for strongly chordal graphs [8].

For cographs equality does not hold. For example,

$$\text{when} \quad G = (P_3 \oplus P_3) \otimes (P_3 \oplus P_3) \quad \text{then} \quad \gamma_{w3}(G) = 4 \quad \text{and} \quad \gamma_{r3}(G) = 6.$$

Let G be a graph and let $k \in \mathbb{N}$. For a function $g : V(G) \to \{0, \dots, k\}$ we write $\|g\|_G = \sum_{x \in V(G)} g(x)$. Furthermore, for $S \subset V(G)$ we write $g(S) = \sum_{x \in S} g(x)$.

Theorem 3. *There exists an $O(k^2 \cdot n)$ algorithm to compute the weak $\{k\}$-domination number for cographs when a cotree is a part of the input.*

Proof. Let $k \in \mathbb{N}$. For a cograph H and $q \in \mathbb{N} \cup \{0\}$, define

$$W(H, q) = \min \{ \|g\|_H \mid g : V(H) \to \{0, \dots, k\} \quad \text{and}$$
$$\forall_{x \in V(G)} \, g(x) = 0 \quad \Rightarrow \quad g(N(x)) + q \geq k \}.$$

When a cograph H is the union of two smaller cographs H_1 and H_2 then

$$\gamma_{wk}(H) = \gamma_{wk}(H_1) + \gamma_{wk}(H_2).$$

In such a case, we have

$$W(H, q) = W(H_1, q) + W(H_2, q).$$

When a cograph H is the join of two cographs H_1 and H_2 then the minimal cost of a weak $\{k\}$-dominating function is bounded from above by $2k$. Then

$$W(H, q) = \min \{ W_1 + W_2 \mid W_1 = W(H_1, q + W_2) \quad \text{and} \quad W_2 = W(H_2, q + W_1) \}.$$

The weak $\{k\}$-domination number of a cograph G, $W(G, 0)$, can be obtained via the above recursion, spending $O(k^2)$ time in each of the n nodes in the cotree. This completes the proof. ∎

Remark 1. Similar results can be obtained for, e.g., the $\{k\}$-domination number [13] and the (j, k)-domination number [26,27].

Remark 2. A frequently studied generalization of cographs is the class of P_4-sparse graphs. A graph is P_4-sparse if every set of 5 vertices induces at most one P_4 [19,22]. We show in the full version that the rainbow domination problem can be solved in linear time on P_4-sparse graphs.

3   Weak $\{k\}$-L-Domination on Trivially Perfect Graphs

Chang et al. were able to solve the k-rainbow domination problem (and the weak $\{k\}$-domination problem) for two subclasses of strongly chordal graphs, namely for trees and for blockgraphs. In order to obtain linear-time algorithms, they introduced a variant, called the weak $\{k\}$-L-domination problem [7,8]. In this section we show that this problem can be solved in $O(k \cdot n)$ time for trivially perfect graphs.

Definition 2. *A $\{k\}$-assignment of a graph G is a map L from $V(G)$ to ordered pairs of elements from $\{0, \ldots, k\}$. Each vertex x is assigned a label $L(x) = (a_x, b_x)$, where a_x and b_x are elements of $\{0, \ldots, k\}$. A weak $\{k\}$-L-dominating function is a function $w : V(G) \rightarrow \{0, \ldots, k\}$ such that, for each vertex x the following two conditions hold.*

$$w(x) \geq a_x, \text{ and}$$
$$w(x) = 0 \Rightarrow w(N[x]) \geq b_x.$$

The weak $\{k\}$-L-domination number is defined as

$$\gamma_{wkL}(G) = \min \{ \|g\| \mid g \text{ is a weak} \{k\}\text{-L-dominating function on } G \}.$$

Notice that

$$\forall_{x \in V(G)} \ L(x) = (0, k) \quad \Rightarrow \quad \gamma_{wk}(G) = \gamma_{wkL}(G).$$

Definition 3. *A graph is trivially perfect if it has no induced P_4 or C_4.*

Wolk investigated the trivially perfect graphs as the comparability graphs of forests. Each component of a trivially perfect graph G has a model which is a rooted tree T with vertex set $V(G)$. Two vertices of G are adjacent if, in T, one lies on the path to the root of the other one. Thus each path from a leaf to the root is a maximal clique in G and these are all the maximal cliques. See [9,17] for the recognition of these graphs. In the following we assume that a rooted tree T as a model for the graph is a part of the (connected) input.

We simplify the problem by using two basic observations. (See [7,8] for similar observations.) Let T be a rooted tree which is the model for a connected trivially perfect graph G. Let R be the root of T; note that this is a universal vertex in G. We assume that G is equipped with a $\{k\}$-assignment L, which attributes each vertex x with a pair (a_x, b_x) of numbers from $\{0, \ldots, k\}$.

(I) There exists a weak $\{k\}$-L-dominating function g of minimal cost such that

$$\forall_{x \in V(G) \setminus \{R\}} \ a_x > 0 \quad \Rightarrow \quad g(x) = a_x.$$

(II) There exists a weak $\{k\}$-L-dominating function g of minimal cost such that

$$\forall_{x \in V(G) \setminus \{R\}} \ a_x = 0 \quad \text{and} \quad b_x \leq \sum_{y \in N[x]} a_y \quad \Rightarrow \quad g(x) = 0.$$

Definition 4. *The reduced instance of the weak $\{k\}$-L-domination problem is the subtree T' of T with vertex set $V(G') \setminus W$, where*

$$W = \{\, x \mid x \in V(G) \setminus \{R\} \quad and \quad a_x > 0 \,\} \;\; \cup$$
$$\{\, x \mid x \in V(G) \setminus \{R\} \quad and \quad a_x = 0 \quad and \quad \sum_{y \in N[x]} a_y \geq b_x \,\}.$$

The labels of the reduced instance are, for $x \neq R$, $L(x) = (a'_x, b'_x)$, where

$$a'_x = 0 \quad and \quad b'_x = b_x - \sum_{y \in N[x]} a_y,$$

and the root R has a label $L(R) = (a'_R, b'_R)$, where

$$a'_R = a_R \quad and \quad b'_R = \max \{\, 0, \, b - \sum_{x \in V(G) \setminus \{R\}} a_x \,\}.$$

The previous observations prove the following lemma.

Lemma 1. *Let T' and L' be a reduced instance of a weak $\{k\}$-L-domination problem. Then*

$$\gamma_{wkL}(G) = \gamma_{wkL'}(G') + \sum_{x \in V(G) \setminus \{R\}} a_x.$$

In the following, let G be a connected, trivially perfect graph and let G be equipped with a $\{k\}$-assignment. Let $G' = (V', E')$ be a reduced instance with model a T' and a root R, and a reduced assignment L'. Let g be a weak $\{k\}$-L'-dominating function on G' of minimal cost. Notice that we may assume that

$$\boxed{\forall_{x \in V(G') \setminus \{R\}} \; g(x) \in \{0, 1\}.}$$

Let x be an internal vertex in the tree T' and let Z be the set of descendants of x. Let P be the path in T' from x to the root R. Assume that Z is a union of r distinct cliques, say $B_1, \ldots, B_r$. Assume that the vertices of each B_j are ordered $x_1^j, \ldots, x_{r_j}^j$ such that

$$\boxed{p \leq q \quad \Rightarrow \quad b'_{x_p^j} \geq b'_{x_q^j}.}$$

Define $d_{x_p^j} = b'_{x_p^j} - p + 1$. Relabel the vertices of Z as $z_1, \ldots, z_\ell$ such that

$$\boxed{p \leq q \quad \Rightarrow d_{z_p} \geq d_{z_q}.}$$

Lemma 2. *There exists an optimal weak $\{k\}$-L'-dominating function g such that $g(z_i) \geq g(z_j)$ when $i < j$.*

We refer to [20] for further details.

Definition 5. *For* $a \in \{0, \ldots, k\}$, $a \geq a'_R$, *let* $\Gamma(G', L', a)$ *be the minimal cost over all weak* $\{k\}$-*L'-dominating functions* g *on* G' *on condition that* $g(P) \geq a$.

Lemma 3. *Define* $d_{z_{\ell+1}} = a$. *Let* $i^* \in \{1, \ldots, \ell+1\}$ *be such that*

(a) $\max \{ a, d_{z_i^*} \} + i^* - 1$ *is smallest possible, and*
(b) i^* *is smallest possible with respect to* (a).

Let $H = G' - Z$. *Let* L^H *be the restriction of* L' *to* $V(H)$ *with the following modifications.*

$$\forall_{y \in P} \ b_y^H = \max \{ 0, \ b'_y - i^* + 1 \}.$$

Let $a^H = \max \{ a, d_{z_{i^*}} \}$. *Then*

$$\Gamma(G', L', a) = \Gamma(H, L^H, a^H) + i^* - 1.$$

We refer to [20] for further details.

The previous lemmas prove the following theorem.

Theorem 4. *Let* G *be a trivially perfect graph with* n *vertices. Let* T *be a rooted tree that represents* G. *Let* $k \in \mathbb{N}$ *and let* L *be a* $\{k\}$-*assignment of* G. *Then there exists an* $O(k \cdot n)$ *algorithm that computes a weak* $\{k\}$-*L-dominating function of* G.

The related (j, k)-domination problem can be solved in linear time on trivially perfect graphs. The weak $\{k\}$-L-domination problem can be solved in linear time on complete bipartite graphs. A detailed discussion can be found in the full version [20].

4   2-Rainbow Domination of Interval Graphs

In [3] the authors ask four questions, the last one of which is, whether there is a polynomial algorithm for the 2-rainbow domination problem on (proper) interval graphs. In this section we show that 2-rainbow domination can be solved in polynomial time on interval graphs.

We use the equivalence of the 2-rainbow domination problem with the weak $\{2\}$-domination problem. The equivalence of the two problems, when restricted to trees and interval graphs, was observed in [3]. Chang et al., proved that it holds for general k when restricted to the class of strongly chordal graphs [8]. The class of interval graphs is properly contained in that of the strongly chordal graphs.

An interval graph has a consecutive clique arrangement. That is a linear ordering $[C_1, \ldots, C_t]$ of the maximal cliques of the interval graph such that, for each vertex, the cliques that contain it occur consecutively in the ordering [16].

Brešar and Šumenjak proved the following theorem.

Theorem 5. (See [3]). *When G is an interval graph,*

$$\gamma_{w2}(G) = \gamma_{r2}(G).$$

In the following, let $G = (V, E)$ be an interval graph.

Lemma 4. *There exists a weak $\{2\}$-dominating function g, with $g(V) = \gamma_{r2}(G)$, such that every maximal clique has at most 2 vertices assigned the value 2.*

Proof. Assume that C_i is a maximal clique in the consecutive clique arrangement of G. Assume that C_i has 3 vertices x, y and z with $g(x) = g(y) = g(z) = 2$. Assume that, among the three of them, x has the most neighbors in $\cup_{j \geq i} C_j$ and that y has the most neighbors in $\cup_{j \leq i} C_j$. Then any neighbor of z is also a neighbor of x or it is a neighbor of y. So, if we redefine $g(z) = 1$, we obtain a weak $\{2\}$-dominating function with value less than $g(V)$, a contradiction. □

Lemma 5. *There exists a weak $\{2\}$-dominating function g with minimum value $g(V) = \gamma_{r2}(G)$ such that every maximal clique has at most four vertices with value 1.*

Proof. The proof is similar to that of Lemma 4. Let C_i be a clique in the consecutive clique arrangement of G. Assume that C_i has 5 vertices x_i, $i \in \{1, \ldots, 5\}$, with $g(x_i) = 1$ for each i. Order the vertices x_i according to their neighborhoods in $\cup_{j \geq i} C_j$ and according to their neighborhoods in $\cup_{j \leq i} C_j$. For simplicity, assume that x_1 and x_2 have the most neighborhoods in the first union of cliques and that x_3 and x_4 have the most neighbors in the second union of cliques. Then $g(x_5)$ can be reduced to zero; any other vertex that has x_5 in its neighborhood already has two other 1's in it.

This proves the lemma. □

Theorem 6. *There exists a polynomial algorithm to compute the 2-rainbow domination number for interval graphs.*

The proof of this theorem and some remarks are moved to the full version [20].

We obtained similar results for the class of permutation graphs. We refer to the full version [20] for the details.

5   NP-Completeness for Splitgraphs

A graph G is a splitgraph if G and $\bar{G}$ are both chordal. A splitgraph has a partition of its vertices into two sets C and I, such that the subgraph induced by C is a clique and the subgraph induced by I is an independent set.

Although the NP-completeness of k-rainbow domination for chordal graphs was established in [7], their proof does not imply the intractability for the class of splitgraphs. However, that is easy to mend.

Theorem 7. *For each $k \in \mathbb{N}$, the k-rainbow domination problem is NP-complete for splitgraphs.*

Proof. Since domination is NP-complete for splitgraphs [4], this proves that k-rainbow domination is NP-complete for $k = 1$. For $k \geq 2$, assume that G is a splitgraph with maximal clique C and independent set I. Construct an auxiliary graph G' by making $k - 1$ pendant vertices adjacent to each vertex of C. Thus G' has $|V(G)| + |C|(k-1)$ vertices, and G' remains a splitgraph. We prove that

$$\gamma_{rk}(G') = \gamma(G) + |C| \cdot (k - 1).$$

We first show that

$$\gamma_{rk}(G') \leq \gamma(G) + |C| \cdot (k - 1).$$

Consider a dominating set D of G with $|D| = \gamma(G)$. We use D to construct a k-rainbow function f for G' as follows:

- For any $v \in D$, if $v \in C$, let $f(v) = [k]$; else, if $v \in I$, let $f(v) = \{k\}$;
- For any $v \in V(G) \setminus D$, let $f(v) = \varnothing$;
- For the $k - 1$ pendant vertices attaching to a vertex $v \in C$, if $f(v) = [k]$, then f assigns to each of these pendant vertices an empty set. Otherwise, if $f(v) = \varnothing$, then f assigns the distinct size-1 sets $\{1\}, \{2\}, \ldots, \{k - 1\}$ to these pendant vertices, respectively.

It is straightforward to check that f is a k-rainbow function. Moreover, we have

$$\gamma_{rk}(G') \leq \sum_{x \in V(G')} |f(x)| = \gamma(G) + |C| \cdot (k - 1).$$

We now show that

$$\gamma_{rk}(G') \geq \gamma(G) + |C| \cdot (k - 1).$$

Consider a minimizing k-rainbow function f for G'. Without loss of generality, we further assume that f assigns either $\varnothing$ or a size-1 subset to each pendant vertex.[1] Define $D \subseteq V(G)$ as

$$D = \{ x \mid f(x) \neq \varnothing \text{ and } x \in V(G) \}.$$

That is, D is formed by removing all the pendant vertices in G', and selecting all those vertices where f assigns a non-empty set. Observe that D is a dominating set of G.[2] Moreover, we have

---

[1] Otherwise, if a pendant vertex p attaching v is assigned a set with two or more labels, say $f(p) = \{\ell_1, \ell_2, \ldots\}$, we modify f into f' so that $f'(p) = \{\ell_1\}$, $f'(v) = f(v) \cup (f(p) \setminus \{\ell_1\})$, and $f'(x) = f(x)$ for the remaining vertices; the resulting f' is still a minimizing k-rainbow function.

[2] That is so because for any $v \in V(G) \setminus D$, we have $f(v) = \varnothing$ so that the union of labels of v's neighbor in G' is $[k]$; however, at most $k - 1$ neighbors of v are removed, and each was assigned a size-1 set, so that v must have at least one neighbor in D.

$$|D| = \sum_{x \in C} [f(x) \neq \varnothing] + \sum_{x \in I} [f(x) \neq \varnothing]$$

$$\leq \sum_{x \in V(G') \setminus I} |f(x)| - |C| \cdot (k-1) + \sum_{x \in I} |f(x)|$$

$$\leq \sum_{x \in V(G')} |f(x)| - |C| \cdot (k-1),$$

where the first inequality follows from the fact that for each $v \in C$ and its corresponding pendant vertices P_v,

$$|f(v)| + \sum_{x \in P_v} |f(x)| - (k-1) = \begin{cases} 0 & \text{if } f(v) = \varnothing \\ \geq 1 & \text{if } f(v) \neq \varnothing. \end{cases}$$

Consequently, we have

$$\gamma(G) \leq |D| \leq \gamma_{rk}(G') - |C| \cdot (k-1).$$

This proves the theorem. $\square$

Similarly, we have the following theorem.

Theorem 8. *For each $k \in \mathbb{N}$, the weak $\{k\}$-domination problem is NP-complete for splitgraphs.*

Proof. Let G be a splitgraph with maximal clique C and independent set I. Construct the graph G' as in Theorem 7, by adding $k-1$ pendant vertices to each vertex of the maximal clique C. We prove that

$$\gamma_{wk}(G') = \gamma(G) + |C| \cdot (k-1).$$

First, let us prove that

$$\gamma_{wk}(G') \leq \gamma(G) + |C|(k-1).$$

Let D be a minimum dominating set. Construct a weak $\{k\}$-domination function $g : V(G') \to \{0, \dots, k\}$ as follows.

(i) For $x \in D \cap C$, let $g(x) = k$.
(ii) For $x \in D \cap I$, let $g(x) = 1$.
(iii) For $x \in V(G) \setminus D$, let $g(x) = 0$.
(iv) For a pendant vertex x with $N(x) \in D$, let $g(x) = 0$.
(v) For a pendant vertex x with $N(x) \notin D$, let $g(x) = 1$.

It is easy to check that g is a weak $\{k\}$-dominating function with cost

$$\gamma_{wk}(G') \leq \sum_{x \in V(G')} g(x) = \gamma(G) + |C| \cdot (k-1).$$

To prove the converse, let g be a weak $\{k\}$-dominating function for G' of minimal cost. We may assume that $g(x) \in \{0,1\}$ for every pendant vertex x. Define

$$D = \{\, x \mid x \in V(G) \quad \text{and} \quad g(x) > 0 \,\}.$$

Then D is a dominating set of G. Furthermore,

$$
\begin{aligned}
\gamma(G) \leq |D| = \sum_{x \in C} [g(x) > 0] &+ \sum_{x \in I} [g(x) > 0] \\
&\leq \sum_{x \in V(G') \setminus I} g(x) - |C| \cdot (k-1) + \sum_{x \in I} g(x) \\
&\leq \sum_{x \in V(G')} g(x) - |C| \cdot (k-1) \\
&\leq \gamma_{wk}(G') - |C| \cdot (k-1).
\end{aligned}
$$

This proves the theorem. □

Acknowledgments. The authors would like to thank the anonymous reviewers for helpful comments. Wing-Kai Hon and Hsiang-Hsuan Liu were supported in part by MOST grant 102-2221-E-007-068. Hung-Lung Wang was supported in part by MOST grant 103-2221-E-141-004. Both grants are from the Ministry of Science and Technology, Taiwan.

References

1. Aharoni, R., Szabó, T.: Vizing's conjecture for chordal graphs. Discrete Math. **309**, 1766–1768 (2009)
2. Brešar, B., Henning, M., Rall, D.: Rainbow domination in graphs. Taiwanese J. Math. **12**(1), 213–225 (2008)
3. Brešar, B., Šumenjak, T.: On 2-rainbow domination in graphs. Discrete Appl. Math. **155**(17), 2394–2400 (2007)
4. Bertossi, A.: Dominating sets for split and bipartite graphs. Inf. Process. Lett. **19**, 37–40 (1984)
5. Bretscher, A., Corneil, D., Habib, M., Paul, C.: A simple linear time LexBFS cograph recognition algorithm. SIAM J. Discrete Math. **22**, 1277–1296 (2008)
6. Cameron, P.: Two-graphs and trees. Discrete Math. **127**, 63–74 (1994)
7. Chang, G., Wu, J., Zhu, X.: Rainbow domination on trees. Discrete Appl. Math. **158**, 8–12 (2010)
8. Chang, G., Li, B., Wu, J.: Rainbow domination and related problems on strongly chordal graphs. Discrete Appl. Math. **161**, 1395–1401 (2013)
9. Chu, F.: A simple linear time certifying LBFS-based algorithm for recognizing trivially perfect graphs and their complements. Inf. Process. Lett. **107**, 7–12 (2008)
10. Corneil, D., Lerchs, H., Stewart-Burlingham, L.: Complement reducible graphs. Discrete Appl. Math. **3**, 163–174 (1981)
11. Corneil, D., Perl, Y., Stewart, L.: A linear recognition algorithm for cographs. SIAM J. Comput. **14**, 926–934 (1985)

12. Courcelle, B.: The expression of graph properties and graph transformations in monadic second-order logic. Handbook of Graph Grammars and Graph Transformations. World Scientific Publishing Co. Inc., River Edge (1997)

13. Domke, G., Hedetniemi, S., Laskar, R., Fricke, G.: Relationships between integer and fractional parameters of graphs. In: Alavi, Y., Chartrand, G., Oellermann, O., Schwenk, A. (eds.) Graph Theory, Combinatorics, and Applications: Proceedings of the 6th Quadrennial International Conference on the Theory and Applications of Graphs 1 (Kalamzaoo 1988), pp. 371–387. Wiley (1991)

14. Farber, M.: Domination, independent domination, and duality in strongly chordal graphs. Discrete Appl. Math. **7**, 115–130 (1984)

15. Gallai, T.: Transitiv orientierbare graphen. Acta Math. Acad. Sci. Hung. **18**, 25–66 (1967). A translation appears in Ramírez-Alfonsín, J., Reed, B. (eds.): Perfect Graphs. Interscience series in discrete mathematics and optimization. John Wiley & Sons, Chichester (2001)

16. Gilmore, P., Hoffman, A.: A characterization of comparability graphs and of interval graphs. Canadian J. Math. **16**, 539–548 (1964)

17. Golumbic, M.: Trivially perfect graphs. Discrete Math. **24**, 105–107 (1978)

18. Habib, M., Paul, C.: A simple linear time algorithm for cograph recognition. Discrete Appl. Math. **145**, 183–197 (2005)

19. Hoàng, C.: A class of perfect graphs, Master's Thesis. School of Computer Science, McGill University, Montreal (1983)

20. Hon, W.-K., Kloks, T., Liu, H.-H., Wang, H.-L.: Rainbow domination and related problems on some classes of perfect graphs (2015). arXiv:1502.07492 [cs.DM]

21. Hartnell, B., Rall, D.: On dominating the cartesian product of a graph and K_2. Discussiones Mathematicae Graph Theory **24**, 389–402 (2004)

22. Jamison, B., Olariu, S.: A tree representation for P_4-sparse graphs. Discrete Appl. Math. **35**, 115–129 (1992)

23. Pai, K., Chiu, W.: 3-Rainbow domination number in graphs. In: Proceedings of the Institute of Industrial Engineers Asian Conference 2013, pp. 713–720. Springer, Science+Business Media Singapore (2013)

24. van Rooij, J., Exact exponential-time algorithms for domination problems in graphs, Ph.D. Thesis, Utrecht University (2011)

25. van Rooij, J., Bodlaender, H.: Exact algorithms for dominating set. Discrete Appl. Math. **159**, 2147–2164 (2011)

26. Rubalcaba, R., Slater, P.: Efficient (j, k)-domination. Discussiones Mathematicae Graph Theory **27**, 409–423 (2007)

27. Rubalcaba, R., Slater, P.: A note on obtaining k dominating sets from a k-dominating function on a tree. Bull. Inst. Combin. Appl. **51**, 47–54 (2007)

28. Šumenjak, T., Rall, D., Tepeh, A.: Rainbow domination in the lexicographic product of graphs (2012). arXiv:1210.0514

29. Vizing, V.: Some unsolved problems in graph theory. Uspehi Mat. Naukno. (in Russian) **23**, 117–134 (1968)

30. Yen, C.: 2-Rainbow domination and its practical variations on weighted graphs. In: Chang, R.-S., Jain, L.C., Peng, S.-L. (eds.) Advances in Intelligent Systems and Applications - Volume 1. Smart Innovation, Systems and Technologies, vol. 20, pp. 59–68. Springer, Berlin (2013)

Efficient Computation of Generalized Ising Polynomials on Graphs with Fixed Clique-Width

Tomer Kotek[1]([⊠]) and Johann A. Makowsky[2]

[1] TU Vienna, Vienna, Austria
tkotek@tuwien.ac.at
[2] Technion — Israel Institute of Technology, Haifa, Israel
janos@cs.technion.ac.il

Abstract. Graph polynomials which are definable in Monadic Second Order Logic (MSOL) on the vocabulary of graphs are Fixed-Parameter Tractable (FPT) with respect to clique-width. In contrast, graph polynomials which are definable in MSOL on the vocabulary of hypergraphs are fixed-parameter tractable with respect to tree-width, but not necessarily with respect to clique-width. No algorithmic meta-theorem is known for the computation of graph polynomials definable in MSOL on the vocabulary of hypergraphs with respect to clique-width. We define an infinite class of such graph polynomials extending the class of graph polynomials definable in MSOL on the vocabulary of graphs and prove that they are Fixed-Parameter Polynomial Time (FPPT or XP) computable, i.e. that they can be computed in time $O(n^{f(k)})$, where n is the number of vertices and k is the clique-width.

1 Introduction

In recent years there has been growing interest in graph polynomials, functions from graphs to polynomial rings which are invariant under isomorphism. Graph polynomials encode information about the graphs in a compact way in their evaluations, coefficients, degree and roots. Therefore, efficient computation of graph polynomials has received considerable attention in the literature. Since most graph polynomials which naturally arise are ♯P-hard to compute (see e.g. [11,25,40]), a natural perspective under which to study the complexity of graph polynomials is that of *parameterized complexity*.

Parameterized complexity is a successful approach to tackling NP-hard problems [18,20], by measuring complexity with respect to an additional *parameter* of the input; we will be interested in the parameters tree-width and clique-width. A computational problem is **fixed-parameter tractable** (FPT) with respect to a parameter k if it can be solved in time $f(k) \cdot p(n)$, where f is a

Tomer Kotek was supported by the Austrian National Research Network S11403-N23 (RiSE) of the Austrian Science Fund (FWF) and by the Vienna Science and Technology Fund (WWTF) grant PROSEED.

M.T. Hajiaghayi and M.R. Mousavi (Eds.): TTCS 2015, LNCS 9541, pp. 135–146, 2016.
DOI: 10.1007/978-3-319-28678-5_10

computable function of k, n is the size of the input, and $p(n)$ is a polynomial in n. Many NP-hard problems are fixed-parameter tractable for an appropriate choice of parameter, see [20] for many examples. Every problem in the infinite class of decision problems definable in *Monadic Second Order Logic* (MSOL) is fixed-parameter tractable with respect to tree-width by Courcelle's Theorem [9,13,14] (though the result originally was not phrased in terms of parameterized complexity).

The computation problem we consider for a graph polynomial $P(G; x_1, \ldots, x_r)$ is the following:

P – Coefficients

Instance : A graph G

Problem : Compute the coefficients $a_{i_1,\ldots,i_r}$ of the monomials $x_1^{i_1} \cdots x_r^{i_r}$.

For graph polynomials, a parameterized complexity theory with respect to tree-width has been developed. Here, the goal is to compute, given an input graph, the table of coefficients of the graph polynomial. The Tutte polynomial has been shown to be fixed-parameter tractable [8,37]. [34] used a logical method to study the parameterized complexity of an infinite class of graph polynomials, including the Tutte polynomial, the matching polynomial, the independence polynomial and the Ising polynomial. [34] showed that the class of graph polynomials definable in MSOL in the vocabulary of hypergraphs[1] is fixed-parameter tractable. This class contains the vast majority of graph polynomials which are of interest in the literature.

Going beyond tree-width to clique-width the situation becomes more complicated. [15] studied the class of graph polynomials definable in MSOL *in the vocabulary of graphs*. They proved that every graph polynomial in this class is fixed-parameter tractable with respect to clique-width. However, this class of graph polynomials does not contain important examples such as the chromatic polynomial, the Tutte polynomial and the matching polynomial. In fact, [21] proved that the chromatic polynomial and the Tutte polynomial are not fixed-parameter tractable with respect to clique-width (under the widely believed complexity-theoretic assumption that FPT $\neq$ W[1]). [36] proved that the chromatic polynomial and the matching polynomial are **Fixed-Parameter Polynomial Time (FPPT)**[2] computable with respect to clique-width, meaning that they can be computed in time $n^{f(cw(G))}$, where n is the size of the graph, $cw(G)$ is the clique-width of the graph and f is a computable function. [27] proved an analogous result for the Ising polynomial. The main result of this paper is a meta-theorem generalizing the fixed-parameter polynomial time computability of the matching polynomial and the Ising polynomial to an infinite family of graph polynomials definable in MSOL analogous to [15].

---

[1] In [14], MSOL in the vocabulary of hypergraphs is denoted MS_2, while MSOL in the vocabulary of graphs is denoted MS_1.

[2] Note that the class XP of slicewise polynomial-time languages (see [18, Chap. 15]) is exactly the class of FPPT languages.

Theorem 1. *Let P be an* MSOL-*Ising polynomial. P is fixed-parameter polynomial time computable with respect to clique-width.*

The class of MSOL-Ising polynomials is defined in Sect. 2.1.

2 Preliminaries

Let $[k] = \{1, \ldots, k\}$. Let τ_G be the vocabulary of graphs $\tau_G = \langle \mathbf{E} \rangle$ consisting of a single binary relation symbol $\mathbf{E}$. A k-graph is a structure $(V, E, R_1, \ldots, R_k)$ which consists of a simple graph $G = (V, E)$ together with a partition $R_1, \ldots, R_k$ of V. Let τ_G^k denote the vocabulary of k-graphs $\tau_G^k = \langle \mathbf{E}, \mathbf{R}_1, \ldots, \mathbf{R}_k \rangle$ extending τ_G with unary relation symbols $\mathbf{R}_1, \ldots, \mathbf{R}_k$.

The class $CW(k)$ of k-graphs of clique-width at most k is defined inductively. Singletons belong to $CW(k)$, and $CW(k)$ is closed under disjoint union $\sqcup$ and two other operations, $\rho_{i \to j}$ and $\eta_{i,j}$, to be defined next. For any $i, j \in [k]$, $\rho_{i \to j}(G, \bar{R})$ is obtained by relabeling any vertex with label R_i to label R_j. For any $i, j \in [k]$, $\eta_{i,j}(G, \bar{R})$ is obtained by adding all possible edges (u, v) between members of R_i and members of R_j. The clique-width of a graph G is the minimal k such that there exists a labeling $\bar{R}$ for which $(G, \bar{R})$ belongs to $CW(k)$. We denote the clique-width of G by $cw(G)$. The clique-width operations $\rho_{i \to j}$ and $\eta_{i,j}$ are well-defined for k-graphs. The definitions of these operations extend naturally to structures (V, E, S) which expand k-graphs with $S \subseteq v$.

A k-expression is a term t which consists of singletons, disjoint unions $\sqcup$, relabeling $\rho_{i \to j}$ and edge creations $\eta_{i,j}$, which witnesses that the graph $val(\mathsf{t})$ obtained by performing the operations on the singletons is of clique-width at most k. Every graph of tree-width at most k is of clique-width at most $2^{k+1} + 1$, cf. [16]. While computing the clique-width of a graph is NP-hard, S. Oum and P. Seymour showed that given a graph of clique-width k, finding a $(2^{3k+2} - 1)$-expression is fixed parameter tractable with clique-width as parameter, cf. [38,39].

For a formula φ, let $qr(\varphi)$ denote the quantifier rank of φ. For every $q \in \mathbb{N}$ and vocabulary τ, we denote by $\text{MSOL}^q(\tau)$ the set of MSOL-formulas on the vocabulary τ which have quantifier rank at most q. For two τ-structures $\mathcal{A}$ and $\mathcal{B}$, we write $\mathcal{A} \equiv^q \mathcal{B}$ to denote that $\mathcal{A}$ and $\mathcal{B}$ agree on all the sentences of quantifier rank q.

Definition 1 (Smooth Operation). *An ℓ-ary operation* Op *on τ-structures is called smooth if for all $q \in \mathbb{N}$, whenever $\mathcal{A}_j \equiv^q \mathcal{B}_j$ for all $1 \leq j \leq \ell$, we have*

$$\mathsf{Op}(\mathcal{A}_1, \ldots, \mathcal{A}_\ell) \equiv^q \mathsf{Op}(\mathcal{B}_1, \ldots, \mathcal{B}_\ell).$$

Smoothness of the clique-width operations is an important technical tool for us:

Theorem 2 (Smoothness, cf. [33])

1. *For every vocabulary τ, the disjoint union $\sqcup$ of two τ-structures is smooth.*
2. *For every $1 \leq i \neq j \leq k$, $\rho_{i \to j}$ and $\eta_{i,j}$ are smooth.*

It is convenient to reformulate Theorem 2 in terms of *Hintikka sentences* (see [19]):

Proposition 3 (Hintikka Sentences). *Let τ be a vocabulary. For every $q \in \mathbb{N}$ there is a finite set*

$$\mathcal{H}_\tau^q = \{h_1, \ldots, h_\alpha\}$$

of $\mathrm{MSOL}^q(\tau)$-sentences such that

1. *Every $h \in \mathcal{H}_\tau^q$ has a finite model.*
2. *The conjunction $h_1 \wedge h_2$ of any two distinct $h_1, h_2 \in \mathcal{H}_\tau^q$ is unsatisfiable.*
3. *Every $\mathrm{MSOL}^q(\tau)$-sentence θ is equivalent to exactly one finite disjunction of sentences in $\mathcal{H}_\tau^q$.*
4. *Every τ-structure $\mathcal{A}$ satisfies a unique member $hin_\tau^q(\mathcal{A})$ of $\mathcal{H}_\tau^q$.*

In order to simplify notation we omit the subscript τ in hin_τ^q when τ is clear from the context.

Let $\tau_{\mathbf{S}}$ the be the vocabulary consisting of the binary relation symbol $\mathbf{E}$ and the unary relation symbol $\mathbf{S}$. Let $\tau_{\mathbf{S},k}$ extend $\tau_{\mathbf{S}}$ with the unary relation symbols $\mathbf{R}_1, \ldots, \mathbf{R}_k$. From Theorem 2 and Proposition 3 we get:

Theorem 4. *For every $k \in \mathbb{N}^+$:*

1. *There is $\mathsf{F}_\sqcup : \mathcal{H}_{\tau_{\mathbf{S},k}}^q \times \mathcal{H}_{\tau_{\mathbf{S},k}}^q \to \mathcal{H}_{\tau_{\mathbf{S},k}}^q$ such that, for every $\mathcal{M}_1$ and $\mathcal{M}_2$, $\mathsf{F}_\sqcup(hin^q(\mathcal{M}_1), hin^q(\mathcal{M}_2)) = hin^q(\mathcal{M}_1 \sqcup \mathcal{M}_2)$.*
2. *For every unary operation $\mathsf{op} \in \{\rho_{p\to q}, \eta_{p,q} : p, q \in [k]\}$, there is $\mathsf{F}_{\mathsf{op}} : \mathcal{H}_{\tau_{\mathbf{S},k}}^q \to \mathcal{H}_{\tau_{\mathbf{S},k}}^q$ such that, for every $\mathcal{M}$, $\mathsf{F}_{\mathsf{op}}(hin^q(\mathcal{M})) = hin^q(\mathsf{op}(\mathcal{M}_1 \sqcup \mathcal{M}_2))$.*

2.1   MSOL-Ising Polynomials

For every $t \in \mathbb{N}^+$, let $\tau_t = \tau \cup \{\mathbf{S}_1, \ldots, \mathbf{S}_t\}$, where $\mathbf{S}_1, \ldots, \mathbf{S}_t$ are new unary relation symbols.

Definition 5 (MSOL-Ising Polynomials). *For every $t \in \mathbb{N}^+$, $\theta \in \mathrm{MSOL}(\tau_t)$ and $G = (V, E)$ we define $P_{t,\theta}(G; \bar{X}, \bar{Y})$ as follows:*

$$P_{t,\theta}(G; \bar{X}, \bar{Y}) = \sum_{\substack{S_1 \sqcup \cdots \sqcup S_t = V: \\ G \models \theta(S_1, \ldots, S_t)}} \prod_{i=1}^{t} X_i^{|S_i|} \prod_{1 \leq i_1 \leq i_2 \leq t} Y_{i_1, i_2}^{|(S_{i_1} \times S_{i_2}) \cap E|}$$

$P_{t,\theta}$ *is the sum over partitions $S_1, \ldots, S_t$ of V such that $(G, S_1, \ldots, S_t)$ satisfies θ of the monomials obtained as the product of $X_i^{|S_i|}$ for all $1 \leq i \leq t$ and $Y_{i_1, i_2}^{|(S_{i_1} \times S_{i_2}) \cap E|}$ for all $1 \leq i_1 < i_2 \leq t$.*

Example 1 (Ising polynomial). The trivariate Ising polynomial $Z(G; x, y, z)$ is a partition function of the Ising model from statistical mechanics used to study phase transitions in physical systems in the case of constant energies and external field. $Z(G; x, y, z)$ is given by

$$Z(G; x, y, z) = \sum_{S \subseteq V} x^{|S|} y^{|\partial S|} z^{|E(S)|}$$

where ∂S denotes the set of edges between S and $V \backslash S$, and $E(S)$ denotes the set of edges inside S. $Z(G; x, y, z)$ was the focus of study in terms of hardness of approximation in [24] and in terms of hardness of computation under the exponential time hypothesis was studied in [27]. [27] also showed that $Z(G; x, y, z)$ is fixed-parameter polynomial time computable.

$Z(G; x, y, z)$ generalizes a bivariate Ising polynomial, which was studied for its combinatorial properties in [7]. [7] showed that $Z(G; x, y, z)$ contains the matching polynomial, the van der Waerden polynomial, the cut polynomial, and, on regular graphs, the independence polynomial and clique polynomial.

The evaluation of $P_{2,\text{true}}(G; X_1, X_2, Y_{1,1}, Y_{1,2}, Y_{2,2})$ at $X_1 = x, X_2 = 1, Y_{1,1} = z, Y_{1,2} = y$ and $Y_{2,2} = 1$ gives $Z(G; x, y, z)$ and therefore $Z(G; x, y, z)$ is an MSOL-Ising polynomial.

Example 2 (Independence-Ising Polynomial). The independence-Ising polynomial $I_{Is}(G; x, y)$ is given by

$$I_{Is}(G; x, y) = \sum_{\substack{S \subseteq V \\ S \text{ is an independent set}}} x^{|S|} y^{|\partial S|}$$

$I_{Is}(G; x, y)$ contains the independence polynomial as the evaluation $I(G; x) = I_{Is}(G; x, 1)$. See the survey [32] for a bibliography on the independence polynomial. The evaluation $y = 0$ is $I_{Is}(G; x, 0) = (1 + x)^{iso(G)}$, where $iso(G)$ is the number of isolated vertices in G. $I_{Is}(G; x, y)$ is an evaluation of an MSOL-Ising polynomial:

$$I_{Is}(G; x, y) = P_{2,\theta_I}(G; 1, x, 1, y, 1)$$

where $\theta_I(S) = \forall x \forall y \ (E(x, y) \rightarrow (\neg S_2(x) \vee \neg S_2(y)))$.

Example 3 (Dominating-Ising Polynomial). The Dominating-Ising polynomial is given by $D_{Is}(G; x, y, z)$

$$D_{Is}(G; x, y, z) = \sum_{\substack{S \subseteq V \\ S \text{ is a dominating set}}} x^{|S|} y^{|\partial S|} z^{|E(S)|}$$

where ∂S denotes the set of edges between S and $V \backslash S$. $D_{Is}(G; x, y, z)$ contains the domination polynomial $D(G; x)$. $D(G; x)$ is the generating function of its dominating sets and we have $D_{Is}(G; x, 1, 1) = D(G; x)$. The domination polynomial first studied in [10] and it and its variations have received considerable attention in the literature in the last few years, see e.g. [1–6,12,17,26,29,31]. Previous research focused on combinatorial properties such as recurrence relations and location of roots. Hardness of computation was addressed in [30].

$D_{Is}(G; x, y, z)$ encodes the degrees of the vertices of G: the number of vertices with degree j is the coefficient of xy^j in $D_{Is}(G; x, y, z)$. $D_{Is}(G; x, y, z)$ is an MSOL-Ising polynomial given by $P_{2, \theta_D}(G; x, 1, z, y, 1)$, where

$$\theta_D = \forall x \, (S_1(x) \vee \exists y \, (S_1(y) \wedge \mathbf{E}(x, y))) \,.$$

2.2 MSOL-Ising Polynomials vs MSOL-Polynomials

Two classes of graph polynomials which have received attention in the literature are:

1. MSOL-polynomials on the vocabulary of graphs, and
2. MSOL-polynomials on the vocabulary of hypergraphs.

See e.g. [28] for the exact definitions. The former class contains graph polynomials such as the independence polynomial and the domination polynomial. The latter class contains graph polynomials such as the Tutte polynomial and the matching polynomial. Every graph polynomial which is MSOL-definable on the vocabulary of graphs is also MSOL-definable on the vocabulary of hypergraphs.

The class of MSOL-Ising polynomials strictly contains the MSOL-polynomials on graphs, see Fig. 1. The containment is by definition. For the strictness, we use the fact that by definition the maximal degree of any indeterminate in an MSOL-polynomial on graphs grows at most linearly in the number of vertices, while the maximal degree of y in the Ising polynomial $Z(K_{n,n}; x, y, z)$ of the complete bipartite graph $K_{n,n}$ equals n^2.

Every MSOL-Ising polynomial $P_{t, \theta}$ is an MSOL-polynomial on the vocabulary of hypergraphs, given e.g. by

$$\sum_{\bar{S}} \sum_{\bar{B}} \prod_{i=1}^{t} X_i^{|S_i|} \prod_{1 \le i_1 \le i_2 \le t} Y_{i_1, i_2}^{|B_{i_1, i_2}|}$$

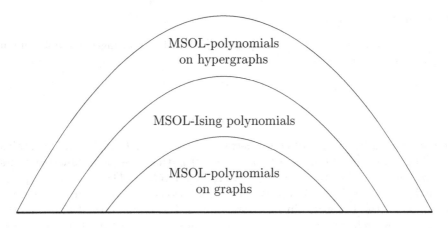

Fig. 1. Containments of classes of graph polynomials definable in MSOL.

where the summation over $\bar{S}$ is exactly as in Definition 5, and the summation over $\bar{B}$ is over tuples $\bar{B} = (B_{i_1,i_2} : 1 \leq i_1 \leq i_2 \leq t)$ of subsets of the edge set of G satisfying $\bigwedge_{i_1,i_2} \psi_{i_1,i_2}$, where

$$\psi_{i_1,i_2} = \forall x \forall y \, (B_{i_1,i_2}(x,y) \leftrightarrow (\mathbf{E}(x,y) \wedge (S_{i_1}(x) \wedge S_{i_2}(y) \vee S_{i_1}(y) \wedge S_{i_2}(x))))$$

We use the fact that $S_1, \ldots, S_t$ is a partition of the set of vertices is definable in MSOL.

3   Main Result

We are now ready to state the main theorem and prove a representative case of it.

Theorem 6 (Main Theorem). *For every MSOL-Ising polynomial $P_{t,\theta}$ there is a function $f(k,\theta,t)$ such that $P_{t,\theta}(G; \bar{X}, \bar{Y}, \bar{Z})$ is computable on graphs G of size n and of clique-width at most k in running time $O(n^{f(k,\theta,t)})$.*

We prove the theorem for graph polynomials of the form

$$Q_\theta(G; X, Y) = \sum_{S : G \models \theta(S)} X^{|S|} Y^{|\partial S|}$$

for every $\theta \in \mathrm{MSOL}(\tau_{\mathbf{S}})$. The summation in Q_θ is over subsets S of the vertex set of G. The graph polynomials Q_θ are a notational variation of $P_{t,\theta}$ with $t = 2$, $X_2 = 1$ and $Y_{1,1} = Y_{2,2} = 1$: for every $\theta \in \mathrm{MSOL}(\tau_2)$, $P_{2,\theta}(G; X, 1, 1, Y, 1) = Q_{\theta'}(G; X, Y)$, where θ' is obtained from θ by substituting $\mathbf{S}_1$ with $\mathbf{S}$ and $\mathbf{S}_2$ with $\neg\mathbf{S}$. The proof for the general case is in similar spirit.

For every $q \in \mathbb{N}$ there is a finite set $\mathfrak{A}_q$ of $\mathrm{MSOL}(\tau_{\mathbf{S},k})$-Ising polynomials such that, for every formula $\theta \in \mathrm{MSOL}^q(\tau_{\mathbf{S}})$, Q_θ is a sum of members of $\mathfrak{A}_q$ (see below). The algorithm computes the values of the members of $\mathfrak{A}_q$ on G by dynamic programming over the parse term of G, and using those values, the value of Q_θ on G.

More precisely, for every $\beta \in \mathcal{H}^q_{\tau_{\mathbf{S},k}}$, let

$$A_\beta(G; \bar{x}, \bar{y}) = \sum_{S : G \models \beta(S)} \prod_{1 \leq c \leq k} x_c^{|S \cap R_c|} \prod_{1 \leq c_1, c_2 \leq k} y_{c_1,c_2}^{|(R_{c_1} \cap S) \times (R_{c_2} \setminus S)|}$$

and let

$$\mathfrak{A}_q = \{A_\beta : \beta \in \mathcal{H}^q_{\tau_{\mathbf{S},k}}\}.$$

Every $\theta \in \mathrm{MSOL}^q(\tau_{\mathbf{S}})$ also belongs to $\mathrm{MSOL}^q(\tau_{\mathbf{S},k})$, and hence there exists by Proposition 3 a set $\mathcal{H} \subseteq \mathcal{H}^q_{\tau_{\mathbf{S},k}}$ such that

$$\theta \equiv \bigvee_{h \in \mathcal{H}} h$$

Hence,

$$Q_\theta(G; X, Y) = \sum_{h \in \mathcal{H}} A_h(G; \bar{x}, \bar{y}) \tag{1}$$

setting $x_c = X$ and $y_{c_1,c_2} = Y$ for all $1 \leq c, c_1, c_2 \leq k$.

For tuples $\bar{b} = ((b_c : c \in [k]), (b_{c_1,c_2} : c_1, c_2 \in [k])) \in [n]^k \times [n]^{k^2}$, let $\mathsf{coeff}_\theta^G(\bar{b}) \in \mathbb{N}$ be the coefficient of

$$\prod_c x_c^{b_c} \prod_{c_1,c_2} y_{c_1,c_2}^{b_{c_1,c_2}}$$

in $A_\beta(G; \bar{x}, \bar{y})$.

Algorithm. Given a k-graph G, the algorithm first computes a parse tree t as in [38,39]. The algorithm then computes $A_\beta(G; \bar{x}, \bar{y})$ for all $\beta \in \mathcal{H}_{\tau_{\mathbf{S}}, k}^q$ by induction over t:

1. If G is a graph of size 1, then $A_\beta(G)$ is computed directly.
2. Let G be the disjoint union of H_A and H_B. We compute $\mathsf{coeff}_\beta^G(\bar{b})$ for every $\beta \in \mathcal{H}_{\tau_{\mathbf{S}},k}^q$ and $\bar{b} \in [n]^k \times [n]^{k^2}$ as follows:

$$\mathsf{coeff}_\beta^G(\bar{b}) = \sum_{h_1,h_2: F_\sqcup(h_1,h_2) \models \beta} \sum_{\bar{d}+\bar{e}=\bar{b}} \mathsf{coeff}_\beta^{H_A}(\bar{d}) \mathsf{coeff}_\beta^{H_B}(\bar{e})$$

3. Let $G = \rho_{p \to q}(H)$. We compute $\mathsf{coeff}_\beta^G(\bar{b})$ for every $\beta \in \mathcal{H}_{\tau_{\mathbf{S}},k}^q$ and $\bar{b} \in [n]^k \times [n]^{k^2}$ as follows:

$$\mathsf{coeff}_\beta^G(\bar{b}) = \sum_{h: F_{\rho_{p \to q}}(h) \models \beta} \sum_{\bar{d}} \mathsf{coeff}_h^H(\bar{d})$$

where the inner summation is over $\bar{d}$ such that

$$b_c = \begin{cases} d_c & c \notin \{p, q\} \\ d_p + d_q & c = q \\ 0 & c = p \end{cases}$$

and

$$b_{c_1,c_2} = \begin{cases} d_{c_1,c_2} & c_1, c_2 \notin \{p, q\} \\ 0 & p \in \{c_1, c_2\} \\ d_{q,q} + d_{p,p} + d_{p,q} + d_{q,p} & c_1 = c_2 = q \\ d_{q,c_2} + d_{p,c_2} & c_1 = q, c_2 \notin \{q, p\} \\ d_{c_1,q} + d_{c_1,p} & c_2 = q, c_1 \notin \{q, p\} \end{cases}$$

4. Let $G = \eta_{p,q}(H)$ with $p \neq q$. Let n_G be the number of vertices in G. We compute $\mathsf{coeff}_\beta^G(\bar{b})$ for every $\beta \in \mathcal{H}_{\tau_{\mathbf{S}},k}^q$ and $\bar{b} \in [n]^k \times [n]^{k^2}$ as follows:

$$\mathsf{coeff}_\beta^G(\bar{b}) = \sum_{h: F_{\eta_{p,q}}(h) \models \beta} \sum_{\bar{d}} \mathsf{coeff}_h^H(\bar{d})$$

where the summation is over $\bar{d}$ such that $b_c = d_c$ and

$$b_{c_1,c_2} = \begin{cases} d_{c_1,c_2} & \{c_1, c_2\} \neq \{p, q\} \\ d_p(n_G - d_q) & c_1 = p, \, c_2 = q \\ d_q(n_G - d_p) & c_1 = q, \, c_2 = p \end{cases}$$

Finally, the algorithm computes Q_θ as the sum from Eq. (1).

3.1 Runtime

The main observations for the runtime analysis are:

- The size of the set $\mathcal{H}^q_{\tau_{S,k}}$ of Hintikka sentences is a function of k but does not depend on n. Let $s^q_{\tau_{S,k}} = |\mathcal{H}^q_{\tau_{S,k}}|$.
- By definition of A_β, for a monomial $\prod_{1 \leq c \leq k} x_c^{i_c} \prod_{1 \leq c_1, c_2 \leq k} y_{c_1,c_2}^{j_{c_1,c_2}}$ to have a non-zero coefficient, it must hold that $i_c \leq n$ and $j_{c_1,c_2} \leq \binom{n}{2}$, since i_c and j_{c_1,c_2} are sizes of sets of vertices and sets of edges, respectively.
- The coefficient of any monomial of A_β is at most 2^n.
- The parse tree guaranteed in [38,39] is of size $O(n^c f_1(k))$ for suitable f_1 and c.

The algorithm performs a single operation for every node of the parse tree.

Singletons: the coefficients of every $A_\beta \in \mathfrak{A}_q$ for a singleton k-graph can be computed in time $O(k)$, which can be bounded by $O(n^k)$.

Disjoint Union, Recoloring and Edge Additions: the algorithm sums over (1) $h \in \mathcal{H}^q_{\tau_{S,k}}$ or pairs $(h_1, h_2) \in \left(\mathcal{H}^q_{\tau_{S,k}} \right)^2$ and (2) over $\bar{d} \in [n]^k \times [n]^{k^2}$ or pairs $(\bar{d}, \bar{e}) \in \left([n]^k \times [n]^{k^2} \right)^2$, then (3) performs a fixed number of arithmetic operations on numbers which can be written in $O(n)$ space. Each node in the parse tree requires time at most $O\left(n^k (s^q_{\tau_{S,k}})^2 \left([n]^k \times [n]^{k^2} \right)^2 \right)$. Since the size of the parse tree is $O(n^c f_1(k))$, the algorithm runs in fixed-parameter polynomial time.

4   Conclusion

We have defined a new class of graph polynomials, the MSOL-Ising polynomials, extending the MSOL-polynomials on the vocabulary of graphs and have shown that every MSOL-Ising polynomial can be computed in fixed-parameter polynomial time. This result raises the question of which graph polynomials are MSOL-Ising polynomials. In previous work [23,28,35] we have developed a method based on connection matrices to show that graph polynomials are not definable in MSOL over either the vocabulary of graphs or hypergraphs.

Problem 1. How can connection matrices be used to show that graph polynomials are not MSOL-Ising polynomials?

The Tutte polynomial does not seem to be an MSOL-Ising polynomial. [22] proved that the Tutte polynomial can be computed in subexponential time for graphs of bounded clique-width. More precisely, the time bound in [22] is of the form $\exp(n^{1-f(cw(G))})$, where $0 < f(i) < 1$ for all $i \in \mathbb{N}$.

Problem 2. Is there a natural infinite class of graph polynomials definable in MSOL which includes the Tutte polynomial such that membership in this class implies *fixed parameter subexponential time* computability with respect to clique-width (i.e., that the graph polynomial is computable in $\exp(n^{1-g(cw(G))})$ time for some function g satisfying $0 < g(i) < 1$ for all $i \in \mathbb{N}$)?

Acknowledgement. We are grateful to Nadia Labai for her comments and suggestions.

References

1. Akbari, S., Alikhani, S., Oboudi, M.R., Peng, Y.-H.: On the zeros of domination polynomial of a graph. Comb. Graphs **531**, 109–115 (2010)
2. Akbari, S., Alikhani, S., Peng, Y.: Characterization of graphs using domination polynomials. Eur. J. Comb. **31**(7), 1714–1724 (2010)
3. Akbari, S., Oboudi, M.R.: Cycles are determined by their domination polynomials. Ars Comb. **116**, 353–358 (2014)
4. Alaeiyan, M., Bahrami, A., Farahani, M.R.: Cyclically domination polynomial of molecular graph of some nanotubes. Digest J. Nanomaterials Biostructures **6**(1), 143–147 (2011)
5. Alikhani, S., Peng, Y.-H.: Dominating sets and domination polynomials of paths. Int. J. Math. Math. Sci. **2009**, 10 (2009)
6. Alikhani, S., Peng, Y.: Introduction to domination polynomial of a graph. Ars Comb. **114**, 257–266 (2014)
7. Andrén, D., Markström, K.: The bivariate ising polynomial of a graph. Discrete Appl. Math. **157**(11), 2515–2524 (2009)
8. Andrzejak, A.: An algorithm for the Tutte polynomials of graphs of bounded treewidth. DMATH: Discrete Math. **190**, 39–54 (1998)
9. Arnborg, S., Lagergren, J., Seese, D.: Easy problems for tree-decomposable graphs. J. Algorithms **12**(2), 308–340 (1991)
10. Arocha, J.L., Llano, B.: Mean value for the matching and dominating polynomial. Discussiones Math. Graph Theor. **20**(1), 57–69 (2000)
11. Bläser, M., Hoffmann, C.: On the complexity of the interlace polynomial. In: STACS 2008, pp. 97–108. IBFI Schloss Dagstuhl (2008)
12. Brown, J.I., Tufts, J.: On the roots of domination polynomials. Graphs Comb. **30**(3), 527–547 (2014)
13. Courcelle, B.: The monadic second-order logic of graphs. i. recognizable sets of finite graphs. Inform. Comput. **85**(1), 12–75 (1990)
14. Courcelle, B., Engelfriet, J.: Graph Structure and Monadic Second-Order Logic: A Language-Theoretic Approach, vol. 138. Cambridge University Press, Cambridge (2012)
15. Courcelle, B., Makowsky, J.A., Rotics, U.: Linear time solvable optimization problems on graphs of bounded clique-width. Theor. Comput. Syst. **33**(2), 125–150 (2000)

16. Courcelle, B., Olariu, S.: Upper bounds to the clique width of graphs. Discrete Appl. Math. **101**(1), 77–114 (2000)
17. Dod, M., Kotek, T., Preen, J., Tittmann, P.: Bipartition polynomials, the ising model and domination in graphs. Discussiones Math. Graph Theor. **35**(2), 335–353 (2015)
18. Downey, R.G., Fellows, M.R.: Parameterized Complexity, vol. 3. springer, Heidelberg (1999)
19. Ebbinghaus, H.-D., Flum, J.: Finite Model Theory. Springer Science & Business Media, Heidelberg (2005)
20. Flum, J., Grohe, M.: Parameterized Complexity Theory. Texts in Theoretical Computer Science. An EATCS Series, vol. 14. Springer, Heidelberg (2006)
21. Fomin, F.V., Golovach, P.A., Lokshtanov, D., Saurabh, S.: Algorithmic lower bounds for problems parameterized with clique-width. In: Proceedings of the Twenty-First Annual ACM-SIAM Symposium on Discrete Algorithms, SODA 2010, Austin, Texas, USA, January 17–19, 2010, pp. 493–502 (2010)
22. Giménez, O., Hlinený, P., Noy, M.: Computing the Tutte polynomial on graphs of bounded clique-width. SIAM J. Discrete Math. **20**(4), 932–946 (2006)
23. Godlin, B., Kotek, T., Makowsky, J.A.: Evaluations of graph polynomials. In: Broersma, H., Erlebach, T., Friedetzky, T., Paulusma, D. (eds.) WG 2008. LNCS, vol. 5344, pp. 183–194. Springer, Heidelberg (2008)
24. Goldberg, L.A., Jerrum, M., Paterson, M.: The computational complexity of two-state spin systems. Random Struct. Algorithms **23**(2), 133–154 (2003)
25. Jaeger, F., Vertigan, D.L., Welsh, D.J.A.: On the computational complexity of the jones and Tutte polynomials. Math. Proc. Cambridge Philos. Soc. **108**(01), 35–53 (1990)
26. Kahat, S.S., Khalaf, A.J.M., Roslan, R.: Dominating sets and domination polynomial of wheels. Asian J. Appl. Sci. 2(3) (2014)
27. Kotek, T.: Complexity of ising polynomials. Comb. Prob. Comput. **21**(05), 743–772 (2012)
28. Kotek, T., Makowsky, J.A.: Connection matrices and the definability of graph parameters. Logical Methods in Computer Science, 10(4) (2014)
29. Kotek, T., Preen, J., Simon, F., Tittmann, P., Trinks, M.: Recurrence relations and splitting formulas for the domination polynomial. Electr. J. Comb. **19**(3), P47 (2012)
30. Kotek, T., Preen, J., Tittmann, P.: Domination polynomials of graph products. To appear in Journal of Combinatorial Mathematics and Combinatorial Computing
31. Kotek, T., Preen, J., Tittmann, P.: Subset-sum representations of domination polynomials. Graphs Comb. **30**(3), 647–660 (2014)
32. Levit, V.E., Mandrescu, E.: The independence polynomial of a graph-a survey. In: Proceedings of the 1st International Conference on Algebraic Informatics, pp. 233–254 (2005)
33. Makowsky, J.A.: Algorithmic uses of the Feferman-Vaught theorem. Ann. Pure Appl. Logic **126**(1), 159–213 (2004)
34. Makowsky, J.A.: Coloured Tutte polynomials and Kauffman brackets for graphs of bounded tree width. Discrete Appl. Math. **145**(2), 276–290 (2005)
35. Makowsky, J.A.: Connection matrices for MSOL-definable structural invariants. In: Ramanujam, R., Sarukkai, S. (eds.) Logic and Its Applications. LNCS (LNAI), vol. 5378, pp. 51–64. Springer, Heidelberg (2009)
36. Makowsky, Johann A., Rotics, Udi, Averbouch, Ilya, Godlin, Benny: Computing graph polynomials on graphs of bounded clique-width. In: Fomin, Fedor V. (ed.) WG 2006. LNCS, vol. 4271, pp. 191–204. Springer, Heidelberg (2006)

37. Noble, S.D.: Evaluating the Tutte polynomial for graphs of bounded tree-width. In: Combinatorics, Probability and Computing, vol. 7, Cambridge University Press (1998)
38. Oum, S.: Approximating rank-width and clique-width quickly. In: Kratsch, D. (ed.) WG 2005. LNCS, vol. 3787, pp. 49–58. Springer, Heidelberg (2005)
39. Oum, S.-I., Seymour, P.: Approximating clique-width and branch-width. J. Comb. Theor. Ser. B **96**, 514–528 (2006)
40. Valiant, L.G.: The complexity of enumeration and reliability problems. SIAM J. Comput. **8**(3), 410–421 (1979)

Infinite Subgame Perfect Equilibrium in the Hausdorff Difference Hierarchy

Stéphane Le Roux$^{(\boxtimes)}$

Université Libre de Bruxelles, Brussels, Belgium
stephane.le.roux@ulb.ac.be

Abstract. Subgame perfect equilibria are specific Nash equilibria in perfect information games in extensive form. They are important because they relate to the rationality of the players. They always exist in infinite games with continuous real-valued payoffs, but may fail to exist even in simple games with slightly discontinuous payoffs. This article considers only games whose outcome functions are measurable in the Hausdorff difference hierarchy of the open sets (*i.e.* Δ_2^0 when in the Baire space), and it characterizes the families of linear preferences such that every game using these preferences has a subgame perfect equilibrium: the preferences without infinite ascending chains (of course), and such that for all players a and b and outcomes x, y, z we have $\neg(z <_a y <_a x \wedge x <_b z <_b y)$. Moreover at each node of the game, the equilibrium constructed for the proof is Pareto-optimal among all the outcomes occurring in the subgame. Additional results for non-linear preferences are presented.

Keywords: Infinite multi-player games in extensive form · Subgame perfection · Borel hierarchy · Preference characterization · Pareto-optimality

1 Introduction

Game theory is the theory of competitive interactions between agents having different interests. Until the late 1960's an agent would usually represent a human or group of humans, when game theory was mainly meant for economics and political science. Then game theory was also applied to evolutionary biology [24] and to theoretical computer science [4], especially to system verification and system synthesis (against given specifications). Classically, the verification or synthesis problem is represented as a game with two players: the system trying to win the game by meeting the specifications, and the environment trying to win the game by preventing the system from doing so. The two players play this game in turn and deterministically on a finite or infinite directed graph, and the key notion is that of winning strategy. For a decade, though, computer scientists such as Ummels [26] have been considering multi-player games to represent more complex verification or synthesis problems, *e.g.* relating to distributed systems.

© IFIP International Federation for Information Processing 2016
Published by Springer International Publishing Switzerland 2016. All Rights Reserved.
M.T. Hajiaghayi and M.R. Mousavi (Eds.): TTCS 2015, LNCS 9541, pp. 147–163, 2016.
DOI: 10.1007/978-3-319-28678-5_11

The notion of winning strategy is specific to two-player win-lose games, but in a multi-player setting it may be replaced with a faithful extension, namely the famous notion of (pure) Nash equilibrium. It does not only accommodate more than two players, it also allows for refined/quantitative objectives.

The deterministic turn-based games on graphs may be unfolded, usually without much loss of information, into deterministic turn-based games on finite or infinite trees, which have been widely studied in game theory. It is one reason why this article focuses on perfect information games in extensive form (*i.e.* played on trees) and their deterministic strategies, unless otherwise stated.

Kuhn [12] proved the existence of Nash equilibrium (NE) in finite games with real-valued payoffs. His proofs uses backward induction and constructs a special kind of NE that was later called subgame perfect equilibrium (SPE) by Selten [23]. An extension of Kuhn's result [13] characterizes the preferences that always yield finite games with NE/SPE: the acyclic preferences. Also, Escardó and Oliva [5] studied generalizations of backward induction in possibly infinite yet well-founded game trees, *i.e.* without infinite plays. The SPE have nice extra properties and are usually preferred over the more general NE: for psychology in a broad sense an SPE amounts to the absence of empty threats, and for system engineering in a broad sense it amounts to stability of a system regardless of the initial state.

The concept of infinite horizon is convenient in economics and also in computer science, *e.g.*, for liveness. Gale and Stewart [9] studied infinite two-player win-lose games, but backward induction is no longer applicable since there may not be leaves to start the induction at. Nevertheless, they proved that if the winning set of each player is open or closed (with the usual topology), one player has a wining strategy. This result was extended by Wolfe [27] for Σ_2^0 and Π_2^0 sets, then by other people to more complex sets, and eventually by Martin for Borel [19] and even quasi-Borel [20] sets. This is called (quasi-)Borel determinacy.

Mertens and Neymann [21, p.1567] found that Borel determinacy can be used to show existence of ϵ-NE in infinite games with bounded Borel-measurable real-valued payoffs. By generalizing their technique, an abstract result about point-classes and determinacy [14] implies a characterization of the preferences that always yield games with NE, in games with (quasi-)Borel-measurable outcome functions, countably many outcomes, and an arbitrary cardinality of players: the preferences whose inverses are well-founded. Then it was shown [16] that two-player antagonistic games (*i.e.* abstract zero-sum games) with finitely many outcomes and (quasi-)Borel-measurable outcome function have SPE.

When the outcome function is a continuous real-valued payoff function, Fudenberg and Levine [8] showed that there is always an SPE in multi-player games. Similar results were obtained recently via an abstract and uniform density argument [16]. The continuity assumption may be slightly relaxed if one is willing to accept approximate SPE. Indeed existence of ϵ-SPE was proved for lower-semicontinuous [6] and upper-semicontinuous [22] payoffs.

However, when the real-valued payoff function is discontinuous enough and the preferences are not antagonistic, there may be no (ϵ-)SPE, as in the following example which is similar to [25, Example 3]. Let a and b be two players with preferences $z <_a y <_a x$ and $x <_b z <_b y$. They are alternatively given the possibility to stop and yield outcomes y and z, respectively, but the outcome is x if no one ever stops.

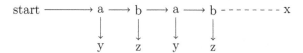

In addition, a real-valued two-player game [7] was recently designed with the following features: it has a similar preference pattern as in the example above, and it has no ϵ-SPE for small enough ϵ even when the players are allowed to use mixed strategies at every node of the game tree. All this shows that Mertens [21] was right when writing that "Subgame perfectness is a completely different issue" from NE. This article solves the issue partially, and the contribution is two-fold. First, it characterizes the linear preferences that always yield SPE in games with outcome functions in the Hausdorff difference hierarchy of the open sets: the preferences that are void of infinite ascending chains (of course) and of the *SPE killer* from the example above. Said otherwise, if $\neg(z <_a y <_a x \wedge x <_b z <_b y)$ holds for all players a and b and all outcomes x, y and z of a multi-player game with outcome function in the difference hierarchy (and preferences without infinite ascending chains), the game has an SPE, and even a *global-Pareto* one, as in Definition 5. Second contribution, the characterization still holds for two-player games with strict weak order preferences, but no longer for three-player games. (Strict weak orders are important since they are an abstraction of the usual preference order over the real-valued payoff functions.)

Section 2 consists of definitions; Sect. 3 proves the characterization for many players and linear preferences; Sect. 4 proves the characterization for two players and strict weak order preferences.

Related Works. Characterizing the preferences that guarantee existence of NE in interesting classes of games is not a new idea: earlier than the two examples above [13,14], Gimbert and Zielonka [10] "characterise the family of payoff mappings for which there always exist optimal positional strategies for both players" in some win-lose games played on finite graphs. Also, [15] characterizes the preferences that guarantee existence of NE in determined, countable two-player game forms.

The notion of SPE has also been studied in connection with system verification and synthesis, at low levels of the Borel hierarchy: in [26] with qualitative objectives, in [2] with quantitative objective for reachability, and in [3] with quantitative objectives and a weak variant of SPE.

Finally, some specific infinite games in extensive form (such as the dollar auction) and especially their SPE have been studied using co-algebraic methods in [1,18].

2   Technical Background

The games in this article are built on infinite trees, which may be defined as prefix-closed sets of finite sequences. The elements of a tree are called nodes. Intuitively, a node represents both a position in the tree and the only path from the root to this position.

Definition 1 (Tree). *Let Σ be a set.*

- Σ^* *(Σ^ω) is the set of finite (infinite) sequences over Σ, and a tree over Σ is a subset T of Σ^* such that $\gamma \sqsubseteq \delta$ (i.e. γ is a prefix of δ) and $\delta \in T$ implies $\gamma \in T$.*
- *For a node γ in a tree T, let $\mathrm{succ}(T, \gamma) := \{\delta \in T \mid \gamma \sqsubseteq \delta \wedge |\delta| = |\gamma| + 1\}$, where $|\gamma|$ is the length of γ.*
- *A tree T is pruned if $\mathrm{succ}(T, \gamma) \neq \emptyset$ for all $\gamma \in T$.*
- *Let T be a tree over Σ. The set $[T]$ is made of the infinite paths of T, namely the elements of Σ^ω whose every finite prefix is in T.*
- *Let T be a tree over Σ. For $\gamma \in T$ let $T_\gamma := \{\delta \in \Sigma^* \mid \gamma\delta \in T\}$.*

In this article the outcomes of a game correspond to some partition of the infinite paths of the game tree, and the subsets of the partition are restricted to the Hausdorff difference hierarchy of the open sets. This hierarchy is defined in, *e.g.*, [11, 22.E], but a probably-folklore result [17, Sect. 2.4] gives an equivalent presentation, which is in turn rephrased in Definition 3 below. These new definitions facilitate the proofs by induction in this article. Then, the Hausdorff-Kuratowski theorem (see, *e.g.*, [11, Theorem 22.27]) implies that, in the Baire space (*i.e.* $\mathbb{N}^{\mathbb{N}}$), the difference hierarchy is equal to Δ_2^0, the sets that are both countable unions of closed sets and countable intersections of open sets. This equality tells how low the difference hierarchy lies in the Borel hierarchy.

For every pruned tree T, let $\{\gamma[T_\gamma] \mid \gamma \in T\}$ be a basis for the open subsets of $[T]$. Definition 2 below is a special case of a more general definition that can be found, *e.g.*, in [11, 22.E].

Definition 2 (Difference Hierarchy). *Let T be a pruned tree, $\theta > 0$ be a countable ordinal, and $(A_\eta)_{\eta < \theta}$ be an increasing sequence of open subsets of $[T]$. $D_\theta((A_\eta)_{\eta < \theta})$ is defined as below.*

$$x \in D_\theta((A_\eta)_{\eta < \theta}) :\Leftrightarrow \quad x \in \cup_{\eta < \theta} A_\eta \text{ and the least } \eta < \theta \text{ with } x \in A_\eta \text{ has parity}$$

opposite to that of θ.

And $D_\theta([T]) := \{D_\theta((A_\eta)_{\eta < \theta}) \mid \forall \eta < \theta, A_\eta \text{ is an open subset of } [T]\}$.

Observation 1. $D_{\theta+1}((A_\eta)_{\eta < \theta+1}) = A_\theta \setminus D_\theta((A_\eta)_{\eta < \theta})$

Definition 3, Lemma 1, and Proposition 1 relate to [17, Sect. 2.4].

Definition 3 (Quasi-Difference Sets). *Let T be a pruned tree. The set $\mathcal{D}([T])$ is defined by transfinite induction below.*

- *Every open set of $[T]$ is in $\mathcal{D}([T])$.*
- *Let $(\gamma_i)_{i \in I}$ be pairwise non-comparable nodes in T, and for all $i \in I$ let $D_i \in \mathcal{D}([T])$ be such that $D_i \subseteq \gamma_i[T_{\gamma_i}]$. Then $\cup_{i \in I} D_i \in \mathcal{D}([T])$.*
- *The complement in $[T]$ of a set in $\mathcal{D}([T])$ is also in $\mathcal{D}([T])$.*

One can prove Observation 2 below by induction along Definition 3.

Observation 2. *Given a node γ of a pruned tree T, and $A \in \mathcal{D}([T])$, $\gamma[T_\gamma] \cap A \in \mathcal{D}([T])$.*

Lemma 1. *$D_\theta \subseteq \mathcal{D}([T])$ for all non-zero countable ordinal θ and all pruned tree T.*

Proof. By induction on θ, which holds for $\theta = 1$ since $D_1([T])$ is made of the open subsets of $[T]$. Let $\theta > 1$ be an ordinal and let $A \in D_\theta$, so $A = D_\theta((A_\eta)_{\eta < \theta})$ for some family $(A_\eta)_{\eta < \theta}$ of open subsets of $[T]$ that is increasing for the inclusion. Every A_η is open so it can be written $\cup_{i \in I} \gamma_{\eta,i}[T_{\gamma_{\eta,i}}]$ where $\gamma_{\eta,i}$ and $\gamma_{\eta,j}$ are not proper prefixes of one another (but are possibly equal) for all $\eta < \theta$ and $i, j \in I$. We can also require some minimality for all $\gamma_{\eta,i}$, more specifically $\gamma \sqsubset \gamma_{\eta,i} \Rightarrow \neg(\gamma[T_\gamma] \subseteq A_\eta)$. Let F consists of the minimal prefixes among $\{\gamma_{\eta,i} \mid i \in I \wedge \eta < \theta\}$, and for all $\gamma \in F$ let $f(\gamma)$ be the least η such that $\gamma = \gamma_{\eta,i}$ for some $i \in I$. So $f(\gamma) < \theta$ for all $\gamma \in F$, and $A \subseteq \cup_{\eta < \theta} A_\eta = \cup_{\gamma \in F} \gamma[T_\gamma]$, so $A = \cup_{\gamma \in F} A \cap \gamma[T_\gamma]$. Let $\gamma \in F$ and let us make a case disjunction to show that $A \cap \gamma[T_\gamma] \in \mathcal{D}([T])$. First case, $f(\gamma)$ and θ have the same parity, so $A \cap \gamma[T_\gamma] = D_{f(\gamma)}((A_\eta \cap \gamma[T_\gamma])_{\eta < f(\gamma)}) \in D_{f(\gamma)}([T])$, so $A \cap \gamma[T_\gamma] \in \mathcal{D}([T])$ by induction hypothesis. Second case, $f(\gamma)$ and θ have opposite parity, so

$$
\begin{aligned}
A \cap \gamma[T_\gamma] &= D_{f(\gamma)+1}((A_\eta \cap \gamma[T_\gamma])_{\eta < f(\gamma)+1}) \text{ by Definition 2,} \\
&= (A_{f(\gamma)} \cap \gamma[T_\gamma]) \setminus D_{f(\gamma)}((A_\eta \cap \gamma[T_\gamma])_{\eta < f(\gamma)}) \text{ by Observation 1,} \\
&= \gamma[T_\gamma] \setminus D_{f(\gamma)}((A_\eta \cap \gamma[T_\gamma])_{\eta < f(\gamma)}) \text{ by definition of } F \ni \gamma \text{ and } f(\gamma), \\
&\in \mathcal{D}([T]) \text{ by induction hypothesis and Observation 2.}
\end{aligned}
$$

Therefore $A \in \mathcal{D}([T])$.

Proposition 1 below shows that the quasi-difference sets coincide with the sets in the difference hierarchy for countable trees, just like quasi-Borel sets [20] and Borel sets coincide on Polish spaces.

Proposition 1. *$\mathcal{D}([T]) = \cup_{\theta < \omega_1} D_\theta([T])$ for all countable pruned tree T.*

Proof. $\cup_{\theta < \omega_1} D_\theta \subseteq \mathcal{D}([T])$ was already proved in Lemma 1, so let $A \in \mathcal{D}([T])$. Let us prove that $A \in D_\theta([T])$ for some θ, by induction on the definition of $\mathcal{D}([T])$. Base case, if A is open, $A \in D_1([T])$.

Second case, let $(\gamma_i)_{i \in I}$ be pairwise non-comparable nodes in T, and let $A_i \in \mathcal{D}([T]) \cap \gamma_i[T_{\gamma_i}]$ for all $i \in I$, such that $A = \cup_{i \in I} A_i$. By induction hypothesis $A_i \in D_{\theta_i}$ for some $\theta_i < \omega_1$. Let $\theta := \sup_{i \in I} \theta_i$, so $\theta < \omega_1$ by countability of T, and $A_i \in D_\theta$ for all $i \in I$. For all $i \in I$ let $(A_{i,\eta})_{\eta<\theta}$ be an increasing sequence of open sets such that $A_i = D_\theta((A_{i,\eta})_{\eta<\theta})$. So $A = \cup_{i\in I}D_\theta((A_{i,\eta})_{\eta<\theta}) = D_\theta((\cup_{i\in I}A_{i,\eta})_{\eta<\theta})$, which shows that $A \in D_\theta([T])$.

Third case, A is the complement of $B \in D_\theta([T])$ for some $\theta < \omega_1$. So $B = D_\theta((B_\eta)_{\eta<\theta})$ for some increasing sequence of open sets of $[T]$. Let $B_\theta = [T]$, so $A = D_{\theta+1}((B_\eta)_{\eta<\theta+1})$.

Informally, a play of a game starts at the root of an infinite game tree and at each stage of the game the unique owner of the current node chooses a child of the node. The articles [14,16] used game trees of the form C^* because it was more convenient and done without much loss of generality, but this article works with general pruned trees because they will be cut in a non-uniform way. Moreover pruned trees are general enough, since leaves in infinite games can be simulated by the pseudo-leaves defined below.

Definition 4 (Game, Subgame, Pseudo-Leaf). *An infinite game g is a tuple $\langle A, T, d, O, v, (\prec_a)_{a\in A}\rangle$ complying with the following.*

- *A is a non-empty set (of players).*
- *T is a non-empty pruned tree (of possible finite plays).*
- *$d : T \to A$ (assigns a decision maker to each stage of the game).*
- *O is a non-empty set (of possible outcomes of the game).*
- *$v : [T] \to O$ (uses outcomes to value the infinite plays in the tree).*
- *Each $\prec_a$ is a binary relation over O (modelling the preference of player a).*

For $\gamma \in T$, the subgame g_γ is defined by $d\gamma : T_\gamma \to A$ such that $d_\gamma(\delta) := d(\gamma\delta)$ and by $v_\gamma : [T_\gamma] \to O$ such that $v_\gamma(p) := v(\gamma p)$. A x-pseudo-leaf of g is a shortest node $\gamma \in T$ such that only the outcome x occurs in g_γ.

Definition 5 (strategy Profile, Induced Play, Global-Pareto Equilibrium). *Let $g = \langle A, T, d, O, v, (\prec_a)_{a\in A}\rangle$ be a game.*

- *A strategy profile is a function $s : T \to T$ such that $s(\gamma) \in \mathrm{succ}(T, \gamma)$ for all $\gamma \in T$. Let S_g be the set of the strategy profiles for g. For $\gamma \in T$ and $s \in S_g$ the subprofile $s_\gamma : T_\gamma \to T_\gamma$ is defined by the equality $\gamma s_\gamma(\delta) = s(\gamma\delta)$.*
- *For $\gamma \in T$ and $s \in S_g$, the play $p = p^\gamma(s)$ induced by s at γ is defined inductively by $p_0 \ldots p_{|\gamma|-1} := \gamma$ and $p_n := s(p_0 \ldots p_{n-1})$ for all $n > |\gamma|$.*
- *A Nash equilibrium is a profile $s \in S_g$ such that*

$$NE_g(s) := \forall s' \in S_g, \forall a \in A, \neg(vop^\epsilon(s) \prec_a vop^\epsilon(s'))$$
$$\wedge \ (\forall \gamma \in T, s(\gamma) \neq s'(\gamma) \Rightarrow d(\gamma) = a))$$

A subgame perfect equilibrium is a profile $s \in S_g$ such that $NE_{g_\gamma}(s_\gamma)$ for all $\gamma \in T$.
- *Let $O' \subseteq O$. One says that $x \in O'$ is Pareto-optimal in O' if for all $y \in O'$ and $a \in A$ such that $x \prec_a y$ there exists $b \in B$ such that $y \prec_b x$. A global-Pareto Nash equilibrium (GP-NE) is an NE whose induced outcome is Pareto-optimal in the outcomes occurring in the underlying game. A GP-SPE is a profile that induces a GP-NE in every subgame.*

The proofs in this article do not build SPE by mere backward induction, but more generally by recursively refining rational behavioral promises. At each stage the refinement is optimal given the existing promises and regardless of the future ones. Since a promise not to choose a specific successor of a given node cannot be represent by a strategy profile, the more general notion of quasi-profile is defined below.

Definition 6 (Quasi Profile). *Let $g = \langle A, T, d, O, v, (\prec_a)_{a \in A} \rangle$ be a game.*

- *A quasi profile is a multivalued function $q : T \multimap T$ such that $\emptyset \neq q(\gamma) \subseteq \text{succ}(\gamma)$ for all $\gamma \in T$. Let Q_g be the set of the quasi profiles for g. For $\gamma \in T$ and $q \in Q_g$ the sub-quasi-profile $q_\gamma : T_\gamma \multimap T_\gamma$ is defined by the equality $\gamma q_\gamma(\delta) = q(\gamma \delta)$.*
- *For $\gamma \in T$ and $q \in Q_g$, the tree induced by q starting at γ is defined inductively by $\epsilon \in T_\gamma(q)$, where ϵ is the empty sequence, and $\delta \in T_\gamma(q) \Rightarrow q_\gamma(\delta) \subseteq T_\gamma(q)$.*
- *Let $q \in Q_g$. Let $(\gamma_i)_{i \in I}$ be the nodes of T such that $\gamma_i \notin q(\gamma)$ for all $\gamma \in T$, and let $G(g, q) := \{g_{\gamma_i} \upharpoonright_{T_{\gamma_i}(q)} \}_{i \in I}$.*

Making a promise in a game g by defining a quasi-profile q splits the game into "smaller" games, formally *via* $G(g, q)$. If the promise is rational, these "smaller" games can be processed independently since gluing any of their respective SPE will yield an SPE for g. Towards this, Observation 3 below suggests that the recursive refinement will lead to a fully defined strategy profile of g, if performed a sufficiently great (ordinal) number of times.

Observation 3. *Let g be a game on a tree T, let q be a quasi profile for g, and let $G(g, q) = \{g_{\gamma_i} \upharpoonright_{T_{\gamma_i}(q)} \}_{i \in I}$. Then $\{\gamma_i T_{\gamma_i}(q)\}_{i \in I}$ is a partition of T.*

3   Many Players with Linearly Ordered Preferences

This section characterizes the linear preferences that always yield SPE in games with $\mathcal{D}_{\omega_1}$-measurable outcome functions: the families of preferences without infinite ascending chains and without the SPE killer, *i.e.* the pattern $z <_a y <_a x \wedge x <_b z <_b y$ for some players a and b and outcomes x, y, z. The main difficulty is tackled by Lemma 2 and corollary 1 below. It consists in slightly generalizing an existing result [16] stating that two-player Borel games with antagonist preferences have SPE by considering preferences that are almost antagonist, but in addition there is an outcome y that is the worst one for both players, and the set of plays with outcome y is a closed set (union an open set). This is then generalized for a set in $\mathcal{D}_{\omega_1}$ by induction, and eventually to multi-player games without the SPE killer thanks to a combinatorial result.

Lemma 2. *Let a game involve two players a and b, preferences $y <_a x_n <_a \cdots <_a x_1$ and $y <_b x_1 <_b \cdots <_b x_n$ for some n, such that all plays without pseudo-leaves have outcome y. The game has a global-Pareto subgame perfect equilibrium.*

Proof. Let us consider only infinite games involving two players a and b, preferences $y <_a x_n <_a \cdots <_a x_1$ and $y <_b x_1 <_b \cdots <_b x_n$ for some n. Let us call a game *weak-stop* if every play without pseudo-leaves has outcome y, and *strong-stop* if in addition every node that does not lie on a play with outcome y has a prefix that is a pseudo-leaf. Note the following: in every weak-stop game the plays with outcome y form a closed set; for every quasi profile q for a weak-stop game g, the set $G(g, q)$ contains only weak-stop games; and modifying the outcome function of a weak-stop game such that it is constant on given subgames yields a weak-stop game. (But the same does not hold for strong-stop games.) Let us call a node of a strong-stop game an a_k (b_k) node if it is owned by player a (b), and if it is the parent of a x_k-pseudo-leaf. Let us call every a_k (b_k) node a a-stop (b-stop) node, and furthermore let us call every a-stop or b-stop node a stop node.

Let us prove the claim by induction on n, which holds for $n = 0$ and $n = 1$, so let us assume that $1 < n$. Five transformations on games are defined below, and they are meant to be applied recursively to a weak-stop game.

1. "Weak-stop towards strong-stop": Let g be a weak-stop game on tree T. Let γ be a node such that g_γ involves more than one outcome but not y. By construction g_γ is an antagonist game, and it amounts, when seeing a pseudo-leaf as a leaf, to a game without infinite plays, but possibly without uniform bound on the length of the plays. By [5] it has a GP-SPE s_γ nonetheless, which induces some x_k. Let us derive a weak-stop game g' from g by modification of the outcome function: for all $p \in [T]$ let $v'(p) := x_k$ if $\gamma \sqsubseteq p$ and $v'(p) := v(p)$ otherwise. Pasting s_γ at node γ on a GP-SPE s' for g' yields a GP-SPE s for g. Formally $s(\gamma\delta) := s_\gamma(\delta)$ for all $\delta \in T_\gamma$ and $s(\delta) := s'(\delta)$ for all $\delta \in T$ such that $\gamma \not\sqsubseteq \delta$.

2. "Emptying the interior of y": Let g be a strong-stop game on tree T and let γ be a y-pseudo-leaf. If $\gamma = \epsilon$ all profiles for g are GP-SPE. If γ is not the root of g let us define a quasi profile q for g by letting the owner of the parent of γ ignore γ. Formally, $q(\delta) := \mathrm{succ}(T, \delta)\backslash\{\gamma\}$ for all $\delta \in T$. Since y is the worst outcome for both players, none will have an incentive to deviate from this promise, regardless of the future choices at the other nodes. $G(g, q)$ contains g_γ and a weak-stop game g'. Combining any profile s_γ for g_γ and a GP-SPE for g' yields a GP-SPE for g.

3. "b chooses x_n": Let g be a strong-stop game on tree T, let γ be a b_n node and let $\delta \in \mathrm{succ}(T, \gamma)$ be an x_n-pseudo-leaf. Let us define a quasi profile q for g by letting b choose δ at γ. Formally, $q(\gamma) := \{\delta\}$ and $q(\alpha) := \mathrm{succ}(T, \alpha)$ for all $\alpha \in T\backslash\{\gamma\}$. Since x_n is b's preferred outcome, she will have no incentive to deviate from this choice, regardless of the choices at the other nodes. So, finding a GP-SPE for every weak-stop game in $G(g, q)$ will complete the definition of a GP-SPE for g.

4. "a ignores x_n": Let g be a strong-stop game. Let γ be a node in g such that g_γ involves outcome y but no b-stop nodes, and such that every subgame of g_γ involving outcome y has an a_k node for some $k < n$. Let us define a quasi profile q for g by letting a ignore all x_n-pseudo-leaves at all a_n nodes below γ. The set $G(g, q)$ is made of games involving only outcome x_n and of one g'

such that g'_γ does not involve x_n. By induction hypothesis g'_γ has an GP-SPE s'_γ, which induces some x_k. Let us define g'' by modification of the outcome function of g: for all $p \in \gamma[T_\gamma]$ let $v''(p) := x_k$ and for all $p \in [T]\backslash\gamma[T_\gamma]$ let $v''(p) := v(p)$. Pasting s'_γ on a GP-SPE s'' for g'' yields a GP-SPE for g.

5. "a chooses x_n": Let g be a strong-stop game. Let γ be such that g_γ involves outcome y and every subgame of g_γ involving y has some a_n nodes but no a_k node for all $k < n$. To build for g_γ a GP-SPE s_γ inducing x_n on all of its subprofiles, it suffices, first, to choose arbitrary profiles for the subgames rooted at the x_n-pseudo-leaves of g_γ, and second, to fix consistently paths from each node to an x_n-pseudo-leaf. This second step can be done by letting player a choose an x_n-pseudo-leaf at some node, which defines a quasi profile q, and by repeating it recursively for the games in $G(g_\gamma, q)$. This s_γ is a GP-SPE because every subprofile induces outcome x_n, which is b's preferred outcome, and the only alternative for a in every subprofile is outcome y since all a-stop nodes are a_n nodes. Let us define g' by modification of the outcome function of g: for all $p \in \gamma[T_\gamma]$ let $v'(p) := x_n$ and for all $p \in [T]\backslash\gamma[T_\gamma]$ let $v'(p) := v(p)$. Pasting s_γ on a GP-SPE s' for g' yields a GP-SPE for g.

Given a weak-stop game g, let us apply to it the five transformations above, sequentially, non-deterministically whenever they are applicable, and until none of them is applicable, *i.e.* possibly an ordinal number of times. This yields a set G of strong-stop games (otherwise Transformation 1 could be applied) whose subgames that involve outcome y all have stop nodes (otherwise Transformation 2 could be applied), without b_n nodes (otherwise Transformation 3 could be applied), such that every subgame that involves y but no b-stop nodes has a subgame without a_k nodes for all $k < n$ (otherwise Transformation 4 could be applied), and such that every subgame h' of every game in G has the following property (otherwise Transformation 5 could be applied): if every subgame of h' has a a-stop node, h' has an a_k node for some $k < n$.

Let h' be a subgame of $h \in G$, and that involves y. If h' has no b-stop nodes, combining the properties above shows that all of its subgames have a-stop nodes, so one of them has only a_n nodes, contradiction, so every subgame of h that involves y has a b-stop node.

For every $h \in G$ let us define the quasi profile q by letting a ignore all the x_n-pseudo-leaves. $G(h, q)$ is made of games involving the outcome x_n only and of one h' void of x_n. Since every subgame of h that involves y has a b-stop node, it also holds for h'. By induction hypothesis, h' has a GP-SPE, which is easily extended to a GP-SPE for h, thus completing the definition of a GP-SPE for the original g.

Corollary 1. *Given a game g with two players a and b, a quasi-Borel measurable outcome function, and preferences $y <_a x_n <_a \cdots <_a x_1$ and $y <_b x_1 <_b \cdots <_b x_n$ for some n. If the plays with outcome y form the union of an open set and a closed set, the game has a global-Pareto subgame perfect equilibrium.*

Proof. Let the plays with outcome y be the union $Y = Y_o \cup Y_c$ of an open set and a closed set. Wlog Y_o and Y_c are disjoint. Let us derive g' from g by removing the plays in Y_o. So the plays of g' that do not yield outcome y form an open set, *i.e.* a disjoint union of clopen balls with defined by the prefixes $(\gamma_i)_{i \in I}$. Every game g_{γ_i} is antagonist and quasi-Borel, so it has an SPE s_i by [16]. Let us define g'' by modification of the outcome function of g': every play going through γ_i yields the outcome induced by s_i. This g'' has a GP-SPE s'' by Lemma 2, and together with the s_i it can be used to build a GP-SPE for g.

Let us extend Corollary 1 from open union closed to the difference hierarchy.

Lemma 3. *Let be a game on a tree T, with two players a and b and preferences $y <_a x_n <_a \cdots <_a x_1$ and $y <_b x_1 <_b \cdots <_b x_n$, and let us assume that each set of plays with outcome x_i is quasi-Borel and that Y the set of plays with outcome y is in $\mathcal{D}_{\omega_1}([T])$. Then the game has a global-Pareto subgame perfect equilibrium.*

Proof. By transfinite induction on the level of Y in the difference hierarchy. The base case where Y is open or closed is solved by Corollary 1.

For the inductive case, let us make a case disjunction depending on the last step of the construction of Y. For the union case, $Y = \cup_{i \in \mathbb{N}} \gamma_i Y_i$ for some Y_i that have lower levels than Y in the difference hierarchy, and where the γ_i are not prefixes of one another. By induction hypothesis each g_{γ_i} has a GP-SPE s_i inducing either y or an outcome $x_{k(i)}$. Let us start the construction of a profile s for g by fixing the s_i as the respective subprofiles for the $g \mid_{\gamma_i T_{\gamma_i}}$. Let us define g' by modification of the outcome function of g: let each play going through γ_i yield the outcome induced by s_i. This is a quasi-Borel game and the plays with outcome y form an open set, so it has a GP-SPE s' by Corollary 1, which we use to complete the definition of s. It is easy to check that s is a GP-SPE for g.

For the complementation case, $Y = [T] \backslash ([T] \backslash Y)$, where $[T] \backslash Y$ is equal to $\cup_{i \in \mathbb{N}} \gamma_i X_i$ for some X_i that have lower levels than $[T] \backslash Y$ (and Y) in the difference hierarchy, and where the γ_i are not prefixes of one another. Since all $Y \cap [\gamma_i T_{\gamma_i}] = [\gamma_i T_{\gamma_i}] \backslash \gamma_i X_i$ have lower levels than Y in the difference hierarchy, by induction hypothesis each g_{γ_i} has a GP-SPE s_i inducing either y or an outcome $x_{k(i)}$. Let us start the construction of a profile s for g by fixing the s_i as the respective subprofiles for the $g \mid_{\gamma_i T_{\gamma_i}}$. Let us define g' by modification of the outcome function of g: let each play going through γ_i yield the outcome induced by s_i. This is a quasi-Borel game and the plays with outcome y form the union of an open set and the closed set $[T] \backslash \cup_{i \in I} [\gamma_i T_{\gamma_i}]$, so it has a GP-SPE s' by Corollary 1, which we use to complete the definition of s. It is easy to check that s is a GP-SPE for g.

The combinatorial Lemma 4 below shows that the "local" absence of the SPE-killer amounts to a very simple "global" structure.

Lemma 4. *Let A be a non-empty set and for all $a \in A$ let $<_a$ be a strict linear order over some non-empty set O. The following assertions are equivalent.*

1. $\forall a, b \in A, \forall x, y, z \in O, \neg(z <_a y <_a x \wedge x <_b z <_b y)$.
2. *There exists a partition $\{O_i\}_{i \in I}$ of O and a linear order $<$ over I such that:*
   (a) *$i < j$ implies $x <_a y$ for all $a \in A$ and $x \in O_i$ and $y \in O_j$.*
   (b) *$<_b|_{O_i} = <_a|_{O_i}$ or $<_b|_{O_i} = <_a|_{O_i}^{-1}$ for all $a, b \in A$.*

If 1. and 2. hold, we may also assume that $<_b|_{O_i}=<_a|_{O_i}^{-1}$ is witnessed for all $i \in I$. Also, O_i is always a $<_a$-interval for all $(i,a) \in I \times A$, as implied by 2a.

Proof. $2 \Rightarrow 1$ is straightforward, so let us assume 1. Let $x \sim y$ stand for $\exists a, b \in A, x \leq_a y \leq_b x$, which defines a reflexive and symmetric relation, and note that due to the SPE killer $z <_a y <_a x \wedge x <_b z <_b y$ the following holds: if $x <_a y$ and $y <_b x$, then $x <_a z <_a y$ iff $y <_b z <_b x$. To show that $\sim$ is transitive too, let us assume that $x \sim y \sim z$. If x, y, z are not pairwise distinct, $x \sim z$ follows directly, so let us assume that they are pairwise distinct, so by assumption there exist $a, b, c, d \in A$ such that $y <_a x <_b y <_c z <_d y$. To show that $x \sim z$ there are three cases depending on where z lies with respect to $y <_a x$, all cases invoking the (above-mentioned) forbidden-pattern argument: if $y <_a z <_a x$ then $x <_b z <_b y$, and $x \sim z$ follows; if $y <_a x <_a z$ then $z <_d x <_d y$ and subsequently $y <_c x <_c z$, by invoking twice the forbidden-pattern argument, and $x \sim z$ follows; third case, let us assume that $z <_a y <_a x$. If $x <_b z$ then $x \sim z$ follows, and if $z <_b x$ then $z <_b x <_b y$, so $y <_c x <_c z$, and $x \sim z$ follows. Therefore $\sim$ is an equivalence relation; let $\{O_i\}_{i \in I}$ be the corresponding partition of O.

Now let us show that the $\sim$-classes are $<_a$-intervals for all a, so let $x \sim y$ and $x <_a z <_a y$. By definition of $\sim$, there exists b such that $y <_b x$, in which case $y <_b z <_b x$ by the forbidden-pattern argument, so $x \sim z$ by definition.

Let $x \in O_i$ and $y \in O_j$ be such that $x <_a y$. Since O_i and O_j are intervals, $x' <_a y'$ for all $x' \in O_i$ and $y' \in O_j$. Since $\neg(x' \sim y')$ by assumption, $x' <_b y'$ for all $b \in A$, by definition of $\sim$. In this case defining $i < j$ meets the requirements.

Before proving 2b let us prove that if $x <_a y$ and $y <_b x$ and $z \sim y$, then $z <_a y$ iff $y <_b z$: this is trivial if z equals x or y, so let us assume that $x \neq z \neq y$, and also that $z <_a y$. If $z <_b y$, then $y <_c z$ for some c since $z \sim y$, and wherever x may lie with respect to $y <_c z$, it always yields a SPE killer using $<_a$ or $<_b$, so $y <_b z$. The converse is similar, it follows actually from the application of this partial result using $<_b^{-1}$ and $<_a^{-1}$ instead of $<_a$ and $<_b$.

Now assume that $<_b|_{O_i} \neq <_a|_{O_i}$ for some O_i, so $x <_a y$ and $y <_b x$ for some $x, y \in O_i$. Let $z, t \in O_i$. By the claim just above $z <_a y$ iff $y <_b z$, so by the same claim again $z <_a t$ iff $t <_b z$, which shows that $<_b|_{O_i}=<_a|_{O_i}^{-1}$. This proves the equivalence.

Finally, let us assume that the assertions hold. By definition of $\sim$, if O_i is not a singleton, $x <_a y$ and $y <_b x$ for some $a, b \in A$ and $x, y \in O_i$, so $<_b|_{O_i}=<_a|_{O_i}^{-1}$ is witnessed.

Theorem 1 extends Lemma 3 to many players and more complex preferences.

Theorem 1. *Let g be a quasi-Borel game with players in A, outcomes in O, and linear preferences $<_a$ for all $a \in A$. Let us assume that the inverses of the $<_a$ are well-ordered, and that there exists a partition $\{O_i\}_{i \in I}$ of O and a linear order $<$ of I such that:*

- $i < j$ implies $x <_a y$ for all $a \in A$ and $x \in O_i$ and $y \in O_j$.
- $<_b|_{O_i}=<_a|_{O_i}$ or $<_b|_{O_i}=<_a|_{O_i}^{-1}$ for all $a, b \in A$ and $i \in I$.

Let us further assume that for all $i \in I$ the plays with outcome in $\cup_{j<i} O_j$ form a $\mathcal{D}_{\omega_1}$ set. Then g has a global-Pareto subgame perfect equilibrium.

Proof. By Lemma 4 let us further assume wlog that $<_b|_{O_j} = <_a|_{O_j}^{-1}$ is witnessed for all $j \in I$, so the O_j are finite by well-ordering. Let us build a GP-SPE for g as the limit of a recursive procedure: Let O_i be such that some outcome of O_i occurs in g and such that for all $j > i$ no outcome from O_j occurs in g. Let $(\gamma_k)_{k \in K}$ be the shortest nodes of g such that the outcomes occuring in g_{γ_k} are in $\cup_{j<i} O_j$ only. Let us define a quasi profile q for g by having the γ_k ignored by their parents. $G(g, q)$ consists of the g_{γ_k} and of a game g'. By Lemma 3 there is a GP-SPE s' for g'. (To see this, replace $\cup_{j<i} O_j$ with one single fresh outcome y, i.e. $y \notin O$ and set $y <_a x$ for all $a \in A$ and $x \in O_i$.) Combining s' with GP-SPE for the g_{γ_k} (obtained recursively) yields a GP-SPE for g, since the choices made in g' hold regardless of the choices made in the g_{γ_k}.

Corollary 2. *Let A and O be non-empty finite sets (of players and of outcomes) and for all $a \in A$ let $<_a$ be a linear preference. The following are equivalent.*

1. $\forall a, b \in A, \forall x, y, z \in O, \neg(z <_a y <_a x \wedge x <_b z <_b y)$.
2. *Every $\mathcal{D}_{\omega_1}$-Gale-Stewart game using A, O and the $<_a$ has a GP-SPE.*

Proof. For 1. $\Rightarrow$ 2. invoke Lemma 4 and Theorem 1, and prove 2. $\Rightarrow$ 1. by contraposition with the following folklore example which is detailed, e.g., in [16].

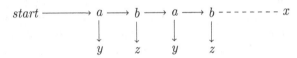

Corollary 2 and the results that lead to it considers linear preference only. Proposition 2 below show that this restriction incurs a loss of generality, which is partly solved in Sect. 4.

Proposition 2. *Let us define two binary relations by $z, t \prec_a x, y$ and $y \prec_b z \prec_b x \prec_b t$.*

1. $\mathcal{D}_{\omega_1}$ *infinite games with players a and b and preferences $\prec_a$ and $\prec_b$ have SPE.*
2. *The SPE killer occurs in any strict linear extensions of $\prec_a$ and $\prec_b$.*

Proof. 1. follows Theorem 2. For 2. let $\prec'_a$ be a linear extension of $\prec_a$. If $x \prec'_a y$ then $z \prec'_a x \prec'_a y$ and $y \prec_b z \prec_b x$. If $y \prec'_a x$ then $t \prec'_a y \prec'_a x$ and $y \prec_b x \prec_b t$.

4   Two Players with Strict Weak Order Preferences

The preferences considered in Proposition 2 are strict weak orders. Informally, strict weak orders are strict partial orders that can be seen as strict linear orders from afar, *i.e.* up to an equivalence relation. Traditionally in game theory the outcomes are real-valued payoff functions $f, g : A \to \mathbb{R}$ and the preferences are defined by $f \prec_a g$ iff $f(a) < g(a)$. These preferences are not strict linear orders but they are strict weak orders, so the results from Sect. 3 are worth generalizing. Strict weak orders are defined below.

Definition 7 (Strict Weak Order). *A strict weak order is a strict partial order whose complement is transitive, i.e. is satisfies $\neg(x \prec x)$ and $x \prec y \wedge y \prec z \Rightarrow x \prec z$ and $\neg(x \prec y) \wedge \neg(y \prec z) \Rightarrow \neg(x \prec z)$.*

Lemma 4 above describes the structure of strict linear orders void of the SPE killer. A similar result for strict weak orders will be useful. Lemma 5 below is part to it, and the other part appears directly in the proof of Lemma 6.

Lemma 5. *Let $\prec_a$ and $\prec_b$ be two strict weak orders over some finite O.*

1. *If $\prec_a$ and $\prec_b$ are void of the SPE killer, if there exists a $\prec_a$-non-extremal element and a $\prec_b$-non-extremal element, and if there is no partition $\{O_u, O_l\}$ of O such that $\neg(x \prec_a y)$ and $\neg(x \prec_b y)$ for all $(x, y) \in O_u \times O_l$, then $\prec_a \cap \prec_b = \emptyset$.*
2. *If $\prec_a \cap \prec_b = \emptyset$, there exists a linear extension $<$ of $\prec_a$ such that $\prec_b \subseteq <^{-1}$.*

Proof.

1. *Let x be $\prec_b$-minimal among the $\prec_a$-maximal elements, and let y be $\prec_a$-minimal among the $\prec_b$-maximal elements, so $x \neq y$ by the partition assumption. Towards a contradiction let us assume that, e.g., x is not $\prec_b$-minimal, and let z be $\prec_b$-minimal. For all $t \prec_b x$, it follows that t is not $\prec_a$-maximal by definition of x, and $\neg(y \prec_a t)$ by absence of the SPE killer. So y is not $\prec_a$-minimal, otherwise z is also $\prec_a$-minimal, thus contradicting the partition assumption. So likewise, for all $t \prec_a y$, it follows that t is not $\prec_b$-maximal by definition of y, and $\neg(x \prec_b t)$ by absence of the SPE killer. So the two-element partition induced by $\{t \in O \mid t \prec_a y \vee t \prec_b x\}$ contradicts the partition assumption. This shows that x is $\prec_b$-minimal and y is $\prec_a$-minimal. Towards a contradiction let us assume that $t(\prec_a \cap \prec_b)z$ for some $t, z \in O$. So $\{x, y\} \cap \{z, t\} = \emptyset$. By the partition assumption z is not both $\prec_a$ and $\prec_b$-maximal, so, e.g., $z \prec_a x$, and t is not both $\prec_a$ and $\prec_b$-minimal. By absence of the SPE killer t is $\prec_b$-minimal, so $y \prec_a t$ by the partition assumption., and subsequently z is $\prec_b$-maximal. By assumption there exists γ that is neither $\prec_b$-maximal nor $\prec_b$-minimal. Wherever γ lies wrt $\prec_a$, the SPE killer occurs.*
2. *By induction on the cardinality of O, which holds for $|O| = 0$. Let x be $\prec_b$-minimal among the $\prec_a$-maximal elements, so x is also $\prec_b$-minimal since $\prec_a \cap \prec_b = \emptyset$. By induction hypothesis let $<_x$ witness the claim for $\prec_a|_{O\setminus\{x\}}$ and $\prec_b|_{O\setminus\{x\}}$. The linear order $< := <_x \cup \{(y, x) \mid y \in O\setminus\{x\}\}$ witnesses the claim.*

Lemma 6 below is a generalization of Corollary 1 from an order-theoretic point of view and a special case thereof from a topological point of view. Due lack of space and strong similarities with the proof of Lemma 2, the proof of Lemma 6 is in appendix.

Lemma 6. *Let g be a game with two players a and b, finitely many outcomes O, and strict weak order preferences void of the SPE killer. If each outcome corresponds to the union of an open set and a closed set, the game has an SPE.*

Furthermore, for every node γ of g let $\{O_1^\gamma, \ldots, O_{n_\gamma}^\gamma\}$ be a partition of the outcomes of g_γ such that $\neg(x \prec_a y)$ and $\neg(x \prec_b y)$ for all $1 \leq k < n_\gamma$ and $(x, y) \in O_{k+1}^\gamma \times O_k^\gamma$. There exists an SPE for g such that the outcome induced at every node γ belongs to $O_{n_\gamma}^\gamma$.

Proof. By induction on the number of outcomes, which holds up to two outcomes. Let us make a case disjunction for the inductive case. First main case, there is no partition $\{O_u, O_l\}$ of O such that $\forall(x, y) \in O_u \times O_l, \neg(x \prec_a y) \wedge \neg(x \prec_b y)$. Let us make a nested case distinction. First nested case, there exists a $\prec_a$-non-extremal element and a $\prec_b$-non-extremal element, so by Lemma 5 let $<$ be a linear extension of $\prec_a$ such that $\prec_b \subseteq <^{-1}$. By [16] the antagonist game with preference $<$ has an SPE, which is also an SPE for $\prec_a$ and $\prec_b$.

Second nested case, one preference, e.g., $\prec_a$ has only extremal elements. By the partition assumption let y be $\prec_b$-maximal and $\prec_a$-minimal. Let Y be the set of plays with outcome y, and let us define a quasi-profile q as follows. Let γ be the parent of a y-pseudo-leaf. If γ is owned by a, let a ignore the y-pseudo-leaves at γ; otherwise let b choose a y-pseudo-leaf at γ. Let us apply this construction recursively (an ordinal number of times) to the games in $G(g, q)$ that do not involve y, until the Y' of each remaining game has empty interior. It is easy to check that Y' is also closed since Y is the union of an open set and a closed set, by assumption. So let $(\gamma_i)_{i \in I}$ be the shortest nodes that are not on any play with outcome y. The g_{γ_i} do not involve y, so by induction hypothesis they have suitable SPE s_i inducing some $x_i \in O$, which allow us to start the definition of a suitable SPE for g. Let us define g' by modification of the outcome function of g: let $v'(\gamma_i p) := x_i$ for all $i \in I$ and let $v'(p) := v(p)$ when $\gamma_i \not\sqsubseteq p$ for all $i \in I$. Let M_a be the $\prec_a$-maximal outcomes, and let us define a quasi-profile q' for g' as follows. Let γ be the parent of a M_a-pseudo-leaf, i.e. a shortest node involving outcome in M_a only. If γ is owned by a, let a choose a M_a-pseudo-leaf at γ; otherwise let b ignore the M_a-pseudo-leave at γ. Let us apply this recursively to the games in $G(g', q)$ that involve y. In the remaining games a can deviate from a play with outcome y only to reach a $(O \backslash) M_a$-pseudo-leaf, so every profile that follows plays with outcome y whenever possible is a suitable SPE.

Second main case, there exist partitions $\{O_u, O_l\}$ of O such that $\forall(x, y) \in O_u \times O_l, \neg(x \prec_a y) \wedge \neg(x \prec_b y)$. Among these partitions let us consider the one with the smallest possible O_u. Let us make a further case disjunction. First case, $|O_l| > 1$. As is now customary, let us start defining a suitable SPE for g by using the induction hypothesis on the maximal subgames involving only outcomes in O_l, and on the game derived from g by replacing outcomes in O_l with a fresh outcome y that is the new $\prec_a$ and $\prec_b$-minimum.

Second case, $|O_l| = \{y\}$. By minimality of $|O_u|$, there is no partition $\{O_{uu}, O_{ul}\}$ of O_u such that $\forall(x, y) \in O_{uu} \times O_{ul}, \neg(x \prec_a y) \wedge \neg(x \prec_b y)$. Therefore the situation is reminiscent of the first main case above, but for $\prec_a|_{O_u}$ and $\prec_b|_{O_u}$ instead of $\prec_a$ and $\prec_b$. In both nested cases from the first main case, there exists some x_n that is, e.g., $\prec_a|_{O_u}$-minimal and $\prec_b|_{O_u}$-maximal. Applying the proof of Lemma 2 almost verbatim yields a suitable SPE for g

Theorem 2. *Let g be a game with two players a and b, finitely many outcomes, a $\mathcal{D}_{\omega_1}$-measurable outcome function, strict weak order preferences such that $\neg(z \prec_a y \prec_a x \wedge x \prec_b z \prec_b y)$ for all outcomes x, y and z. Then the game has an SPE.*

Furthermore, for every node γ of g let $\{O_1^\gamma, \dots, O_{n_\gamma}^\gamma\}$ be a partition of the outcomes of g_γ such that $\neg(x \prec_a y)$ and $\neg(x \prec_b y)$ for all $1 \le k < n_\gamma$ and $(x, y) \in O_{k+1}^\gamma \times O_k^\gamma$. There exists an SPE for g such that the outcome induced at every node γ belongs to $O_{n_\gamma}^\gamma$.

Proof. By induction on the levels in the difference hierarchy of the sets of plays corresponding to the outcomes. The base case holds by Lemma 6.

For the inductive case, let y be an outcome whose corresponding set Y has level more than one in the difference hierarchy, and let us make a case disjunction depending on the last step of the construction of Y. The remainder of the proof can be taken almost verbatim from the proof of Lemma 3, but by replacing "GP-SPE" with "suitable SPE", and by invoking Lemma 6 or the induction hypothesis instead of Corollary 1.

Theorem 2 considers two-player games only. Observation 4 shows that absence of the SPE killer is no longer a sufficient condition for a three-player game with strict weak order preferences to have an SPE.

Observation 4. *Let three players a, b and c have preferences $z \prec_a y \prec_a x$ and $t \prec_b z \prec_b y$ and $x \prec_c t \prec_c y$. (and, e.g., $y \sim_a t$, $z \sim_b x$, and $y \sim_c z$ or $x \sim_c z$)*

1. *The SPE killer does not occur in the strict weak orders $\prec_a$ and $\prec_b$ and $\prec_c$.*
2. *The following game with $\prec_a$ and $\prec_b$ and $\prec_c$ has no SPE.*

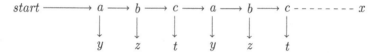

Proof. For 2. Towards a contradiction let us assume that there exists an SPE for the game. Let us consider a node where player a chooses y. Then at the node right above it c chooses to continue to benefit from y, and at the node above b chooses to continue, too. The induced outcome at the node further above is y regardless of the choice of a, and so on up to the root.

Let us make a case disjunction: first case, there exists infinitely many nodes where a chooses y, so b and c always continue by the remark above, so a has an incentive to continue too, to induce outcome x, contradiction. Second case, there exists a node below which a always continues. From then on, one player must stop at some point, otherwise the outcome is x and c has an incentive to stop. The first player to stop cannot be b, otherwise a would stop before b, and it cannot be c, otherwise b would stop before c, contradiction.

Proposition 3 below shows that considering only strict weak orders incurs a loss of generality.

Proposition 3. *Let us define two binary relations by* $\gamma \prec_a y \prec_a x$ *and* $z \prec_a \beta \prec_a \alpha$ *and* $x \prec_b z \prec y$ *and* $\alpha \prec_b \gamma \prec_b \beta$.

1. *The SPE killer occurs in every strict weak order extensions of* $\prec_a$ *and* $\prec_b$.
2. $\mathcal{D}_{\omega_1}$-*games with players* a *and* b *and preferences* $\prec_a$ *and* $\prec_b$ *have SPE.*

Proof.

1. *Let* $\prec_a'$ *be a strict weak order extension of* $\prec_a$. *If* $z \prec_a' y$, *the SPE killer occurs, with* x. *If* $\neg(z \prec_a' y)$, *then* $\gamma \prec_a' \beta$ *and the SPE killer occurs, with* α.
2. *It suffices to prove the claim for games where each outcome set is the union of an open set and a closed set. (Then using a transfinite induction as in the proof of Lemma 3 will do.) The techniques from Lemmas 2 and 6 are suitable here, and used without details. If the outcome* x *does not occur in the game, note that* $\prec_a|_{O\setminus\{x\}}$ *and* $\prec_b|_{O\setminus\{x\}}$ *can be extended into the strict weak orders* $z \prec_a' \gamma \sim_a' \beta \prec_a' y \sim_a' \alpha$ *and* $z \sim_b' \alpha \prec_b' \gamma \prec_b' y \sim_b' \beta$, *respectively, and that the SPE killer is absent from these. So by Theorem 2 the game has an SPE wrt* $\prec_a'$ *and* $\prec_b'$ *and therefore also wrt* $\prec_a$ *and* $\prec_b$. *(And likewise if the outcome* α *does not occur in the game.)*

   Now one can reduce the set for x *to a closed set by letting* a *choose the clopen balls with constant outcome* x *and by letting* b *ignore them. The plays with outcomes different from* x *can be seen as belonging to a union of subgames without* x, *so by the remark above they have SPE. It allows us to replace these subgames with (pseudo)-leaves with outcomes the ones induced by the SPE. Now one can let* a *choose the pseudo-leaves with outcome* α *and* b *ignore them, which yields a game without outcome* α. *So by the remark above there is an SPE for the game.*

Acknowledgements. I thank Vassilios Gregoriades and Arno Pauly for useful discussions. The author is supported by the ERC inVEST (279499) project.

References

1. Abramsky, S., Winschel, V.: Coalgebraic analysis of subgame-perfect equilibria in infinite games without discounting (2012). arXiv preprint
2. Brihaye, T., Bruyère, V., De Pril, J., Gimbert, H.: On subgame perfection in quantitative reachability games. Logical Methods in Computer Science, vol. 9 (2012)
3. Brihaye, T., Bruyère, V., Meunier, N., Raskin, J.F.: Weak subgame perfect equilibria and their application to quantitative reachability (2015)
4. Büchi, J.R., Landweber, L.H.: Solving sequential conditions by finite-state strategies. Trans. Am. Math. Soc. **138**, 295–311 (1969)
5. Escardó, M., Oliva, P.: Selection functions, bar recursion and backward induction. Math. Struct. Comput. Sci. **20**(4), 127–168 (2010)
6. Flesch, J., Kuipers, J., Mashiah-Yaakovi, A., Schoenmakers, G., Solan, E., Vrieze, K.: Perfect-information games with lower-semicontinuous payoffs. Math. Oper. Res. **35**, 742–755 (2010)
7. Flesch, J., Kuipers, J., Mashiah-Yaakovi, A., Schoenmakers, G., Shmaya, E., Solan, E., Vrieze, K.: Non-existence of subgame-perfect ε-equilibrium in perfect information games with infinite horizon. Int. J. Game Theor. **43**(4), 945–951 (2014)

8. Fudenberg, D., Levine, D.: Subgame-perfect equilibria of finite- and infinite-horizon games. J. Econ. Theor. **31**(2), 251–268 (1983)
9. Gale, D., Stewart, F.M.: Infinite games with perfect information. Ann. Math. Stud. **28**, 245–266 (1953)
10. Gimbert, H., Zielonka, W.: Games where you can play optimally without any memory. In: Abadi, M., de Alfaro, L. (eds.) CONCUR 2005. LNCS, vol. 3653, pp. 428–442. Springer, Heidelberg (2005)
11. Kechris, A.S.: Classical Descriptive Set Theory. Graduate Texts in Mathematics, vol. 156. Springer, New York (1995)
12. Kuhn, H.W.: Extensive games and the problem of information. In: Contributions to the Theory of Games II, pp. 193–216 (1953)
13. Le Roux, S.: Acyclic preferences and existence of sequential nash equilibria: a formal and constructive equivalence. In: Berghofer, S., Nipkow, T., Urban, C., Wenzel, M. (eds.) TPHOLs 2009. LNCS, vol. 5674, pp. 293–309. Springer, Heidelberg (2009)
14. Le Roux, S.: Infinite sequential Nash equilibrium. Logical Methods in Computer Science, 9 (2013). Special Issue for the Conference Computability and Complexity in Analysis, (CCA 2011)
15. Le Roux, S.: From winning strategy to Nash equilibrium. Math. Logic Q. **60**, 354–371 (2014)
16. Le Roux, S., Pauly, A.: Infinite sequential games with real-valued payoffs. In: Proceedings of LiCS (2014)
17. Le Roux, S., Pauly, A.: Weihrauch degrees of finding equilibria in sequential games. In: Proceedings of CiE 2015 (2015) (to appear)
18. Lescanne, P., Perrinel, M.: Backward coinduction, Nash equilibrium and the rationality of escalation. Acta Informatica **49**(3), 117–137 (2012)
19. Martin, D.A.: Borel determinacy. Ann. Math. **102**, 363–371 (1975)
20. Martin, D.A.: An extension of Borel determinacy. Ann. Pure Appl. Logic **49**, 279–293 (1990)
21. Mertens, J-F.: Repeated games. In: Proceedings of the International Congress of Mathematicians, pp. 1528–1577. American Mathematical Society (1987)
22. Purves, R., Sudderth, W.: Perfect information games with upper-semicontinuous payoffs. Math. Oper. Res. **36**, 468–473 (2011)
23. Selten, R.: Spieltheoretische Behandlung eines Oligopolmodells mit Nachfrageträgheit. Zeitschrift für die desamte Staatswissenschaft, vol. 121 (1965)
24. Smith, M.J., Price, G.R.: The logic of animal conflicts. Nature **246**, 15–18 (1973)
25. Solan, E., Vieille, N.: Deterministic multi-player dynkin games. J. Math. Econ. **39**(8), 911–929 (2003)
26. Ummels, M.: Rational behaviour and strategy construction in infinite multiplayer games. In: Arun-Kumar, S., Garg, N. (eds.) FSTTCS 2006. LNCS, vol. 4337, pp. 212–223. Springer, Heidelberg (2006)
27. Wolfe, P.: The strict determinateness of certain infinite games. Pac. J. Math. **5**, 841–847 (1955)

Deterministic Algorithm for 1-Median 1-Center Two-Objective Optimization Problem

Vahid Roostapour$^{(\boxtimes)}$, Iman Kiarazm, and Mansoor Davoodi

Institute for Advanced Studies in Basic Sciences (IASBS), Zanjan, Iran
{v.roostapour,i.kiarazm,mdmonfared}@iasbs.ac.ir

Abstract. *k-median* and *k-center* are two well-known problems in facility location which play an important role in operation research, management science, clustering and computational geometry. To the best of our knowledge, although these problems have lots of applications, they have never been studied together simultaneously as a multi objective optimization problem. Multi-objective optimization has been applied in many fields of science where optimal decisions need to be taken in the presence of trade-offs between two or more conflicting objectives. In this paper we consider 1-median and 1-center two-objective optimization problem. We prove that $\Omega(n \log n)$ is a lower bound for proposed problem in one and two dimensions in Manhattan metric. Also, by using the properties of farthest point Voronoi diagram, we present a deterministic algorithm which output the Pareto Front and Pareto Optimal Solutions in $\mathcal{O}(n \log n)$ time.

Keywords: Computational geometry · Pareto optimal solutions · 1-center · 1-median · Multi-objective optimization

1 Introduction

When evaluating different solutions from a design space, it is often the case that more than one criterion comes into play. For example, when choosing a route to drive from one point to another, we may care about the time it takes, the distance traveled and the complexity of the route (e.g. number of turns). When designing a (wired or wireless) network, we may consider its cost, capacity and coverage. Such problems are known as *Multi-Objective Optimization Problems* (MOOP). Multi-objective optimization can be described in mathematical terms as follows:

$$S = \{x \in \mathbb{R}^d : h(x) = 0, g(x) \geq 0\}$$
$$\min [f_1(x), f_2(x), \dots, f_N(x)]$$
$$x \in S,$$

where $N > 1$, f_i is a scalar function for $1 \leq i \leq N$ and S is the set of constraints.

The space in which the objective vector belongs is called *objective space*. The scalar concept of optimality does not apply directly in the multi-objective setting.

© IFIP International Federation for Information Processing 2016
Published by Springer International Publishing Switzerland 2016. All Rights Reserved.
M.T. Hajiaghayi and M.R. Mousavi (Eds.): TTCS 2015, LNCS 9541, pp. 164–178, 2016.
DOI: 10.1007/978-3-319-28678-5_12

Here the notion of *Pareto optimality* and *dominance* has to be introduced. In a multi-objective *minimization* problem, a solution $s_1 \in S$ dominates a solution $s_2 \in S$, denoted by $s_1 \prec s_2$, if $f_i(s_1) \leq f_i(s_2)$ for all $i \in \{1, \ldots, N\}$, with at least one strict inequality. A point s^* is said to be a *Pareto optimum* or a *Pareto optimal solution* for the multi-objective problem if and only if there is no $s \in S$ such that $s \prec s^*$. The image of such an efficient set, i.e., the image of all the efficient solutions in the objective space are called *Pareto optimal front* or *Pareto curve*.

One of the common approaches for such problems is evolutionary algorithms [7]. These algorithms are iterative and converge to Pareto front. However they need more time as the complexity of the Pareto front increases. Moreover, all of these approaches have major problems with local optimums. On the other hand there are some classical approaches like *weighted sum* and *ϵ-constraint* which can apply on MOOPs. Although these approaches guarantee finding solutions on the entire Pareto optimal set for problems having a convex Pareto front, they are largely depend on chosen weight and ϵ vectors respectively. Moreover, these approaches require some information from user about the solution space. Furthermore, in most nonlinear MOOPs, a uniformly distributed set of weight vectors wont necessarily find a uniformly distributed set of Pareto optimal solutions. Also there may exist multiple minimum solutions for a specific weight vector [8]. However we find the Pareto front of a MOOP with deterministic algorithm. Here we consider two famous propounded facility location problems [17].

k-median: In this problem the goal is to minimize summation of distances between each demand point and its nearest center. Charikar et al. proposed the first constant time approximation algorithm which its outputs is $6\frac{2}{3}$ times the optimal [5]. This improved upon the best previously known result of $\mathcal{O}(\log p \log \log p)$, which was obtained by refining and derandomizing a randomized $\mathcal{O}(\log n \log \log n)$-approximation algorithm of Bartal [4]. The currently best known approximation ratio is $3 + \epsilon$ achieved by a local search heuristic of Arya et al. [1]. Moreover, Jain et al. proved that the k-median problem cannot be approximated within a factor strictly less than $1 + 2/e$, unless $\mathsf{NP} \subseteq \mathsf{DTIME}[n^{\mathcal{O}(\log \log n)}]$ [12]. This was an improvement over a lower bound of $1 + 1/e$ [16]. Using sampling technique Meyerson, et al. presented an algorithm with running time $\mathcal{O}(p(\frac{p^2}{\epsilon} \log p)^2 \log(\frac{p}{\epsilon} \log p))$. This was the first k-median algorithm with fully polynomial running time that was independent of n, the size of the data set. It presented a solution that is, with high probability, an $\mathcal{O}(1)$-approximation, if each cluster in some optimal solution has $\Omega(\frac{n \cdot \epsilon}{p})$ points [14]. Har-Peled and Kushal presented a (p, ϵ)-coreset of size $O(p^2/\epsilon^d)$ for k-median clustering of n points in $\mathbb{R}^d$, which its size was independent of n [9]. Also, Har-Peled and Mazumdar showed that there exist small coresets of size $\mathcal{O}(pe^{-d} \log n)$ for the problems of computing k-median clustering for points in low dimension with $(1 + \epsilon)$-approximation. Their algorithm has linear running time for a fixed p and ϵ [10]. Moreover, using random sampling for k-median problem Badoiu et al. proposed a $(1+\epsilon)$-approximation algorithm with $2^{(p/\epsilon)^{\mathcal{O}(1)}} d^{\mathcal{O}(1)} n \log^{\mathcal{O}(p)} n$ expected time [3].

k-center: In this problem the goal is to minimize the maximum distance between each demand point from its nearest center. Megiddo and Supowit proved that k-center and k-median are NP-hard even to approximate the k-center problems sufficiently closely [13]. Hochbaum and Shmoys proposed the first constant factor approximation algorithm which its output is 2 times the optimal. It is the best possible algorithm unless $P \neq NP$ [11]. It is shown that there is an algorithm with $\mathcal{O}(d^{\mathcal{O}(d)}n)$ time for 1-center problem [6]. In the high dimension, Badoiu and Clarkson presented a $(1 + \epsilon)$-approximation algorithm which find a solution in $[2/\epsilon]$ passes using $\mathcal{O}(nd/\epsilon + (1/\epsilon)^5)$ total time and $\mathcal{O}(d/\epsilon)$ space [2]. Also, for problem of 1-center with outliers, Zarrabi-Zadeh and Mukhopadhyay proposed a 2-approximation one pass streaming algorithm in high dimension which for z, as the number of outliers, needs $\mathcal{O}(zd^2)$ space [19]. Moreover, Zarrabi-Zadeh and Chan presented an streaming one pass 3/2-approximation algorithm for 1-center [18]. Badoiu et al. for 1-center problem, extracted a coreset of size $\mathcal{O}(1/\epsilon^2)$ which its solution is $(1 + \epsilon)$-approximation set of points in $\mathbb{R}^d$ [3]. Also, for k-center they presented a $2^{\mathcal{O}((p \log p)/\epsilon^2)}.dn$ time algorithm with $(1 + \epsilon)$-approximation solution using previous result.

1-median and 1-center are practical problems which have not been considered as a two-objective optimization problem yet. Imagine mayor of a small city wants to build a fire station in a way that minimizes the distance between farthest building to the station, also since the number of fire engines is limited and each fire engine must return to the station after a service, it has to minimizes the total distance of station from all other buildings. As an another example, consider power distribution network. Due to the dependency of energy leakage to wire length, minimizing of the longest wire in the network would be regarded as an essential factor. Also, any decrement in total wire length of network considered as a second objective. The first objective is 1-median, $M(u)$, the summation of distances of demand points from center u and the second objective is 1-center, $C(u)$, the farthest input point from center p. It can be described in mathematical terms as follow:

Definition 1. 1-*Median* 1-*Center Two-Objective Optimization Problem:* Let $P = \{p_1, \ldots p_n\}$ be a set of demand points in $\mathbb{R}^d$. Consider functions $M(u) = \sum_{i=1}^n D(u, p_i)$ and $C(u) = \max_{1 \leq i \leq n} D(u, p_i)$ are the values of point $u \in \mathbb{R}^d$ as a center for 1-median and 1-center objectives respectively for a certain distance function D. The goal is finding u^* to minimize the objectives.

We study this problem in one and two dimensions in Manhattan metric. We assume no input points have the same x or y coordinate.

This is a convex combinatorial multi-objective optimization problem which has been studied with a different approach called ϵ-Pareto. In [15] it is shown that this approximate Pareto curve can be constructed in time polynomial in the size of the instance and $1/\epsilon$, but here we propose a deterministic algorithm for computing the exact Pareto curve because of specifying the problem.

This paper starts with considering 1-median and 1-center as two-objective of MOOP in one dimension. We will find the optimal of objectives and in terms of placement of optimums we will also find the Pareto set in time $\mathcal{O}(n)$ (Lemma 1).

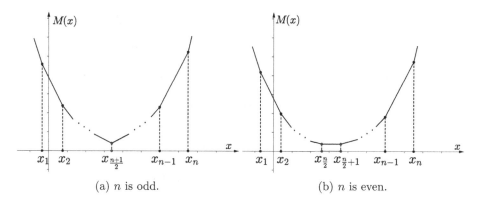

(a) n is odd.   (b) n is even.

Fig. 1. 1-median optimal.

We continue with a proof for convexity of Pareto set. At the end of second section we give an algorithm to compute Pareto optimal front of 1-median 1-center two-objective optimization problem and prove the optimality of the algorithm. In section three the same problem considered in two dimensional space. First we find optimums of 1-median objective. After that by using the properties of farthest point Voronoi diagram we determine the optimum of 1-center. Finally after limiting the solution space to regions which Pareto set lies on, we specifically present Pareto solutions. Convexity of Pareto front is proven in Theorem 2.

2   One Dimensional

Let $P = \{x_1, x_2, \ldots x_n\}$ be a set of input points in one dimension, the goal is to minimize $M(x) = \sum_{i=1}^{n} |x - x_i|$ and $C(x) = \max_{1 \leq i \leq n} |x - x_i|$. According to the properties of the absolute value function and some simple calculations, it is easy to see that $M(x)$ is a continuous piecewise linear function which its minimum depends on n. The minimum can either be one point or an interval which we denote by M_{opt} in the rest of the paper. Also without loss of generality we assume that input points are sorted increasingly. In one dimensional space, $M_{opt} = [m_i, m_j] \subset \mathbb{R}$ for $1 \leq i, j \leq n$ such that $m_i = x_i$, $m_j = x_j$. For odd n we have $j = i$ and for even n, $j = i + 1$. Moreover, the function is strictly decreasing before its minimum and is strictly increasing after it (Fig. 1). For $C(x)$ suppose $c_{opt} \in \mathbb{R}$ denote the point which $C(c_{opt})$ is minimum. Obviously $c_{opt} = (x_1 + x_n)/2$. Similarly to $M(x)$, $C(x)$ is strictly decreasing before optimal point and strictly increasing after that.

Lemma 1. *Pareto optimal set in one dimensional 1-median 1-center two-objective optimization problem is the smallest interval consisting of a solution with 1-center optimal and a solution with 1-median optimal.*

Proof. Suppose that n is even (the proof is similar for odd n). As shown in Fig. 2, there are three different cases:

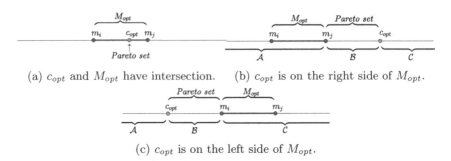

(a) c_{opt} and M_{opt} have intersection.   (b) c_{opt} is on the right side of M_{opt}.

(c) c_{opt} is on the left side of M_{opt}.

Fig. 2. Pareto set computation in one dimension.

First consider the case that c_{opt} and M_{opt} have an intersection (Fig. 2a). In this case the intersection point is the only member of Pareto optimal solutions. Because not only it is optimal in both objectives, but also it is the only point where $C(x)$ is optimal. So it dominates all the other solutions and no solution dominates it.

As shown in Fig. 2b there are three regions in the second case. In region C both functions are strictly increasing. Therefore, c_{opt} has the best value in both objectives. It dominates all solutions of this region. In A, $C(x)$ is strictly decreasing, thus $C(m_j)$ is strictly smaller than 1-center objective of all the other solutions. Moreover, $M(m_j)$ is smaller than or equal with 1-median objective of the other solutions. Hence m_j dominates all solutions of A. Finally we claim that B is Pareto set. By contradiction, suppose it is not true, then there must be a point p which dominates $q \in B$. It has to be on the left side or right side of q. Let p be on the right side, we know that $M(x)$ is strictly increasing in this side. Hence $M(q) < M(p)$ and it contradicts with dominance of p. Similarly there is a contradiction if p lies on the left side of q, because $C(x)$ is strictly decreasing in this side, i.e. $C(q) < C(p)$. This implies that all the solutions that lie on B are Pareto set.

The proof is similar for the third case which c_{opt} is on the left side of M_{opt} (Fig. 2c). □

Lemma 2. *Pareto optimal front of one dimensional 1-median 1-center two-objective optimization problem forms a continuous, convex and piecewise linear function.*

Proof. If there is an intersection between c_{opt} and M_{opt} the lemma is held. Now suppose there is no such intersection and consider c_{opt} is on the right side of M_{opt} (resp. on the left side of M_{opt}). From Lemma 1 for Pareto solutions we have $P_s = [m_j, c_{opt}]$ (resp. $P_s = [c_{opt}, m_i]$). Since $C(x)$ derivation is constant and $M(x)$ is piecewise linear in P_s, the diagram of $M(x)$-$C(x)$ is piecewise linear and break points are $(C(x_i), M(x_i))$ such that $m_j \leq x_i \leq c_{opt}$ (resp. $c_{opt} \leq x_i \leq m_i$). The absolute value of slope of $M(x)$ increases on each linear piece in P_s. Thus

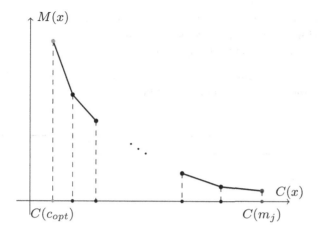

Fig. 3. One dimensional 1-median 1-center two-objective Pareto optimal front.

Pareto optimal front is convex (Fig. 3). Also we can conclude that piecewise linear Pareto front is one-to-one and invertible corresponding to Pareto solutions.   □

Lemma 3. *Computing Pareto front of one dimensional 1-median 1-center two-objective optimization problem requires $\Omega(n \log n)$ time.*

Proof. The proof is based on reduction from *sorting problem*. By contradiction assume there is an algorithm which return set $O = \{(C(\alpha_1), M(\alpha_1)) \cdots, (C(\alpha_m), M(\alpha_m))\}$ –lexicographical ordered break points of the piecewise linear Pareto front function– besides the Pareto solutions interval in $o(n \log n)$ running time. Let $A = \{a_1, \ldots, a_n\}$ is the set of input values of sorting problem, $l = \arg\min_{1 \le i \le n} a_i$ and $h = \arg\max_{1 \le i \le n} a_i$. Suppose $b_1, \ldots, b_{n+1}$ and t are values such that $b_1 < \cdots < b_{n+1} < a_l$ and $t = 2 \cdot a_h - b_1 + 1$, then $B = A \cup \{b_1, \cdots, b_{n+1}, t\}$ is defined in $\mathcal{O}(n)$. For the set B as input points of *one dimensional 1-median 1-center two-objective optimization*, 1-median optimal interval is $[b_{n+1}, a_l]$ and 1-center optimal point is between a_h and t. Using lemma 2 we conclude that $m = n + 1$ and $\alpha_1 = a_l < \cdots < \alpha_{m-1} = a_h < \alpha_m = \frac{(b_1 + t)}{2}$. Therefore, we can sort input points by given algorithm which implies that no algorithms with $o(n \log n)$ running time can compute Pareto front of one dimensional 1-median 1-center two-objective optimization problem.   □

Note 1. If the algorithm output the Pareto optimal front as $O = \{(C(\alpha_1), M(\alpha_1)) - (C(\alpha_2), M(\alpha_2)), \cdots, (C(\alpha_{2m-1}), M(\alpha_{2m-1})) - (C(\alpha_{2m}), M(\alpha_{2m}))\}$, start points and end points of m segments, since the slope of each segment is an integer of $\mathcal{O}(n)$, the segments can be sorted in $\mathcal{O}(n)$. Therefore, we can have sorted break points of Pareto front function and the above proof holds.

Theorem 1. *Algorithm 1 compute one dimensional 1-median 1-center two-objective Pareto front and Pareto solutions interval in $\mathcal{O}(n \cdot \log n)$.*

Proof. $C(x)$ can be computed easily in constant time and $M(x)$ can be computed in $\mathcal{O}(\log n)$ using binary search, we obtain that line 13 is $\mathcal{O}(\log n)$ running time. Therefore, we can conclude that Algorithm 1 is $\mathcal{O}(n \cdot \log n)$.   □

Corollary 1. *Pareto front of one dimensional 1-median 1-center two-objective optimization problem can be computed in $\theta(n \log n)$.*

Algorithm 1. COMPUTE PARETO OPTIMAL FRONT

Input: Set I s.t. $|I| = n$
Output: P_s(Pareto solutions), P_f(Pareto front)
1: Sort I increasingly to $\{x_1, x_2, \ldots, x_n\}$
2: **if** n is even **then**
3:     $b = \frac{n}{2} + 1$
4: **else**
5:     $b = \frac{n+1}{2}$
6: **end if**
7: $P_s = [x_b, (x_1 + x_n)/2]$
8: $P_f = \Phi$
9: Add $\big(C(x_b), M(x_b)\big)$ to P_f
10: $i = b$
11: **while** $x_{i+1} < (x_1 + x_n)/2$ **do**
12:     $i = i + 1$
13:     Add $\big(C(x_i), M(x_i)\big)$ to P_f
14: **end while**
15: Add $\big(C(x_{(x_1+x_n)/2}), M(x_{(x_1+x_n)/2})\big)$ to P_f
16: **return** P_s, P_f

Due to space limitation, Algorithm 1 is just for the case that c_{opt} is on the right side of M_{opt}. The case that c_{opt} is on the left side is similar. If there is an intersection, solution is obviously the intersection point.

3   Two Dimensional

In this section we consider the problem in $\mathbb{R}^2$. The aim is to find the Pareto front and Pareto solutions in terms of M_{opt} and C_{opt}.

3.1   1-Median Objective

For each point $p \in \mathbb{R}^2$ we have:

$$M(u) = \sum_{i=1}^{n} \|u - p_i\|_1$$

$$= \sum_{i=1}^{n} |u_x - p_{ix}| + \sum_{i=1}^{n} |u_y - p_{iy}| \tag{1}$$

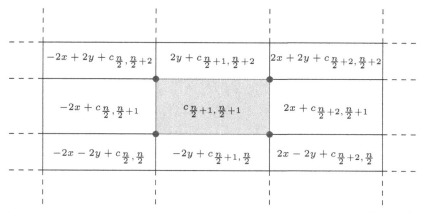

(a) Number of equations is even, 1-median optimal is a rectangular (blue) region.

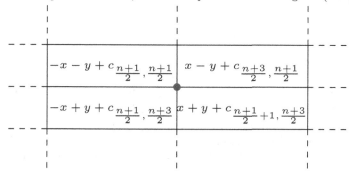

(b) Number of equations is odd, 1-median is a (blue) point.

Fig. 4. 1-median optimal and equation of M(p) in middle cells (Color figure online).

We can observe that we need $\mathcal{O}(n)$ time to deterministically minimize Eq. 1. Moreover, because of the assumption that no points have same coordinate the optimal of M_{opt} may be just a point or area of a rectangle.

In the rest of this paper we assume that n is even (all proofs and discussions are similar when n is odd.). Consider lines $y = p_{i_x}$ and $x = p_{i_y}$ such that $1 \leq i \leq n$ which partition the xy-plane into $(n + 1)^2$ cells where boundary cells are unbounded. The equation of $M(p)$ for points in each cell is the same because of the absolute value function. Furthermore, for points in a column (resp. row) equation of $\sum_{i=1}^{n} |x - p_{i_x}|$ (resp. $\sum_{i=1}^{n} |y - p_{i_y}|$) do not change but for transformation to upper (resp. right) cell coefficient of y (resp. x) increases by 2 (Fig. 4).

3.2   1-Center Objective

Let $\mathcal{FVD}$ be the *farthest point Voronoi diagram* of input points in Manhattan metric, also let $R_{\mathcal{FVD}}(p)$ denote the region of $\mathcal{FVD}$ which consist of p and

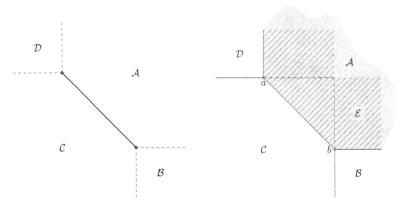

(a) Farthest point Voronoi diagram regions in Manhattan metric.

(b) Possible region for site of C is $A \setminus \mathcal{E}$.

Fig. 5. Farthest point Voronoi diagram properties.

$S_{\mathcal{FVD}}(\mathcal{R})$ denote the site of region $\mathcal{R}$. According to the definition of 1-center objective, $C(p)$ is $\|p - S_{\mathcal{FVD}}(R_{\mathcal{FVD}}(p))\|_1$. Besides the $\mathcal{FVD}$ partition the plane into at least two and at most four regions (Fig. 5a).

According to the structure of $\mathcal{FVD}$, it is impossible for regions A and C to have a common site. However, either B (resp. D) can merge with A (resp. C) or B (resp. D) can merge with C (resp. A), i.e. B and D cannot merge with a common region simultaneously.

Proposition 1. *Site of region C is in $A \setminus \mathcal{E}$. Otherwise distances between points on segment ab and $S_{\mathcal{FVD}}(C)$ are not equal and ab is not an edge of $\mathcal{FVD}$ (Fig. 5b).*

From Proposition 1 it can be concluded that $C(p)$ for $p \in C$ is equal to distance of p from segment ab add up to distance between segment ab and $S_{\mathcal{FVD}}(C)$.

Proposition 2. *As shown in Fig. 6a distances of $m_1, m_2 \in A$ from line ℓ_1 is equal to their distances from segment ab. For point m_1 both distances are obviously the same and are equal to $\|m_1 - m_1'\|_1$. For point m_2 we have $\triangle pqm_2'$ and $\triangle qm_2'a$ as equal isosceles triangles. Therefore, segments qm_2' and qa are equal. Hence $\|m_2 - m_2'\|_1 = \|m_2 - a\|_1$.*

The following two propositions determine the equation of $C(p)$ in the plane and proof that it depends on which region of $\mathcal{FVD}$ includes p.

Proposition 3. *For point $p \in C$ (resp. $p \in A$), $C(p) = k_{opt} + c - p_x - p_y$ (resp. $C(p) = k_{opt} - c + p_x + p_y$) where c is y-intercept of ℓ_1 (Fig. 6b).*

Proof. Suppose equation of line ℓ_1 is $y = -x + c$ and distance between site of C and segment ab is k_{opt}, then projection of point $p = (x, y)$ on ℓ_1 is $p' = (\frac{c-y+x}{2}, \frac{c+y-x}{2})$. Using Propositions 1 and 2 can obtain that:

$$C(p) = k_{opt} + \|p - p'\|_1 = k_{opt} + c - p_x - p_y. \qquad \square$$

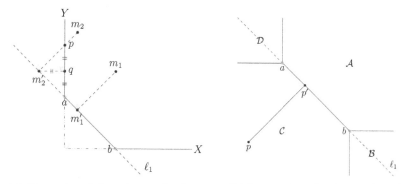

(a) Property of 1-center optimal segment in Manhattan metric.

(b) Projection of points to line ℓ_1.

Fig. 6. Property of 1-center optimal segment in Manhattan metric.

Proposition 4. *In Fig. 7a since site of $\mathcal{D}$ is in hatched region or on its border, for point $q \in \mathcal{D}$ we have $C(q)$ as the distance of point a from $S_{\mathcal{FVD}}(\mathcal{D})$ add up to distance between point a and point q. Also since point a is an $\mathcal{FVD}$ vertex, we know that distance of point a from $S_{\mathcal{FVD}}(\mathcal{D})$ is equal to its distance from $S_{\mathcal{FVD}}(\mathcal{C})$ and as equal to k_{opt}, hence:*

$$C(q) = k_{opt} + \|a - q\|_1 = k_{opt} - c_1 - q_x + q_y$$
$$c_1 = a_y - a_x$$

Similarly it can be proven that for $q \in \mathcal{B}$:

$$C(q) = k_{opt} + \|b - q\|_1 = k_{opt} + c_2 + q_x - q_y$$
$$c_2 = b_y - b_x.$$

Corollary 2. *According to Propositions 3 and 4 we can conclude that points in $\mathcal{A}$ and $\mathcal{C}$ which are on segments parallel to segment ab have the same 1-center objective value. Also for $\mathcal{B}$ and $\mathcal{D}$ these points are on segments perpendicular to ab. Moreover, points on ab are optimal of 1-center objective (Fig. 7b).*

3.3   Pareto Optimal Solutions

Suppose M_{opt} and C_{opt} are calculated. Obviously if they have intersection, it is the set of Pareto solutions. Hence in the rest of this section we assume that M_{opt} and C_{opt} have no intersection.

Possible Region for Pareto Optimal Set. Here the goal is to find the region $\mathcal{P}$ such that its boundary points dominate all points of the plane, i.e. Pareto set is definitely in $\mathcal{P}$.

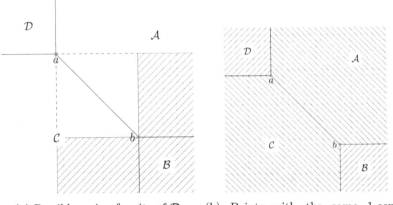

(a) Possible region for site of $\mathcal{D}$.   (b) Points with the same 1-center value.

Fig. 7. Possible region for site of $\mathcal{D}$.

According to the optimal of $M(p)$ and $C(p)$, three cases are possible. In the first case M_{opt} is in regions $\mathcal{A}$ or $\mathcal{C}$, in the second case M_{opt} is in regions $\mathcal{B}$ or $\mathcal{D}$ and in the third case M_{opt} intersects with the axis aligned the edges of $\mathcal{FVD}$. For the first case (Fig. 8a) let e be the lower left point of M_{opt} and let ef and ed be the vertical and horizontal segments hitting the edges of $\mathcal{FVD}$. For all points u on line of ed and w on half-line segment ℓ_1 perpendicular to ℓ_{ed}, $C(u) < C(w)$ and $M(u) \leq M(w)$. Thus u dominates all points on ℓ_1. Similarly for point q on ℓ_{ef} and w on half-line segment ℓ_2, $q \prec w$. There are similar results for other edges of $adefb$ which make us able to conclude that polygon $adefb$ is $\mathcal{P}$.

Second and third cases are similar and we consider them simultaneously (Fig. 8b and c). Let $bcde$ be in region $\mathcal{B}$. Obviously above discussion holds for points p, q, r, s and half-line segments ℓ_1, ℓ_2, ℓ_3 and ℓ_4 respectively. Moreover, a dominates all points of $\mathcal{D}$, any point t on ab dominates all points on horizontal (resp. vertical) half-line segment which starts from t and pass through $\mathcal{C}$ (resp. $\mathcal{A}$) and b dominates all points on ab. Therefore, we can conclude that points on the border of $bcde$ dominate all points outside of it and $bcde$ is $\mathcal{P}$. It is the same when $bcde$ is in $\mathcal{D}$.

Pareto Optimal Solutions. We have shown that $M(p)$ partitions the plane to cells in which equation of $M(p)$ is known. According to this partitioning and C_{opt}, seven cases are possible. First three cases happen when M_{opt} is in $\mathcal{A}$ or $\mathcal{C}$ of $\mathcal{FVD}$. Next three cases occur when M_{opt} is in $\mathcal{B}$ or $\mathcal{D}$. Last case occurs when M_{opt} and axis aligned edges of $\mathcal{FVD}$ have intersection.

The claim is that cells in $\mathcal{P}$ whose equations are $M(p) = \alpha p_x + \beta p_y + c$ such that $\alpha = \beta$, are part of Pareto set.

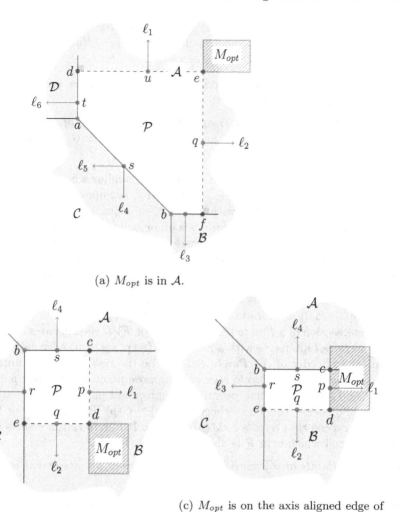

(a) M_{opt} is in $\mathcal{A}$.

(b) M_{opt} is in $\mathcal{B}$.

(c) M_{opt} is on the axis aligned edge of $\mathcal{FVD}$.

Fig. 8. M_{opt} is in $\mathcal{A}$.

Proposition 5. *Let $\mathcal{P}$ be in $\mathcal{A}$. For each cell with $M(p) = \alpha p_x + \beta p_y + c$ where $\alpha/\beta > 1$, points on the right and bottom edges dominate other points of the cell.*

Proof. For each point p in the cell, points with the same 1-median values are on a line which is parallel to $y = -\alpha/\beta$. This line will hit the border of the cell in points p' and p'' such that $p'_x > p''_x$, i.e. p' is on bottom or right edge. Since $\alpha/\beta > 1$ we have $C(p') < C(p) < C(p'')$ and p' dominates p and p''.   □

Proposition 6. *Suppose $\mathcal{P}$ is in $\mathcal{A}$ (resp. $\mathcal{C}$). In $\mathcal{P}$ let q be a point in a cell with equation $M(q) = \alpha q_x + \alpha q_y + c$ such that $\alpha < 0$ (resp. $\alpha > 0$). Suppose ℓ be the* *line passing through q with equation* $y = -x + c'$. *By extending Proposition 5, for all p on ℓ or bellow (resp. on ℓ or above) we have* $M(q) \leq M(p)$. *The same result holds for* $\mathcal{B}$ *and* $\mathcal{D}$ *when ℓ is* $y = x + c'$.

The following lemma introduces special cells in $\mathcal{P}$ which are part of Pareto set. In the rest of this paper we refer to them as *Pareto cells*.

Lemma 4. *Suppose $\mathcal{P}$ is in $\mathcal{A}$ (resp. $\mathcal{C}$). All points like p of cells with $M(p) = \alpha p_x + \beta p_y + c$ such that $\alpha = \beta$ and $\alpha < 0$ (resp. $\alpha > 0$), are all or part of Pareto set.*

Proof. Here we assume $\mathcal{P} \subset \mathcal{A}$ but the proof is similar when $\mathcal{P} \subset \mathcal{C}$. Consider $p \in \mathcal{P}$ such that $M(p) = \alpha p_x + \alpha p_y + c$ and $\alpha < 0$. Suppose q dominates p and ℓ be a line passing through p with $y = -x + c'$ equation. if q is above ℓ then $C(q)$ is greater than $C(p)$. Therefore, q is on or bellow ℓ. By Proposition 6, if q is bellow ℓ it means $M(p) < M(q)$ otherwise q is on ℓ; but if both are in the same cell it concludes that $C(p) = C(q)$ and $M(p) = M(q)$, otherwise $M(p) < M(q)$. We can obtain from these contradictions that no point dominates p.   □

Similar to Lemma 4, cells with $M(p) = \alpha p_x + \beta p_y + c$ such that $\beta = -\alpha$ and $\alpha > 0$ (resp. $\alpha < 0$) are *Pareto cells* in region $\mathcal{B}$ (resp. $\mathcal{D}$).

For intersection of a *Pareto cell* with edges of $\mathcal{FVD}$ several cases are possible. If the *Pareto cell* intersects with a horizontal (resp. vertical) edge, segment from b (resp. a) to border of the *Pareto cell* will be the rest of Pareto solutions, we refer to this segment as *Pareto segment*. Suppose point q dominates $p \in$ *Pareto segment* and let ℓ be the line passing through p and parallel to $y = -x$, then q must be on or below this line, otherwise $C(q) > C(p)$. But if q is on or below ℓ, since p is in a cell that $\alpha/\beta > 1$, $M(p) < M(q)$. If *Pareto cell* intersects with ab, the part of cell which is in $\mathcal{P}$ is also *Pareto cell* (Fig. 9).

Lemma 5. *Points of a Pareto cell in solution space are a segment in objective space.*

Proof. In a *Pareto cell* $M(p) = \alpha p_x + \alpha p_y + c$ ($\alpha < 0$) and $C(p) = p_x + p_y + c'$. Therefore, $M(p) - \alpha C(p) = c''$. This implies that *Pareto cell* in solution space is a segment with $Y - \alpha X = c''$ equation in objective space. It is easy to see that this holds for *Pareto segments*.   □

Theorem 2. *Pareto Front of two dimensional 1-median 1-center two-objective optimization problem is* continuous, convex *and* piecewise linear *function.*

Proof. By Lemma 5 we can conclude that Pareto optimal front is piecewise linear. Since in the sequence of *Pareto cells* from M_{opt} to C_{opt} each cell have a common point with the next cell, the sequence of segments of Pareto front is continuous. Moreover, since in each cell the coordinate of x and y in $M(p)$ is smaller than the previous ones, slope of segment of that cell in objective space will be bigger than segments of previous cells which guarantees convexity of Pareto front.   □

Corollary 3. *Finding Pareto front and Pareto Solution set of two dimensional 1-median 1-center two-objective optimization problem is* $\theta(n \log n)$.

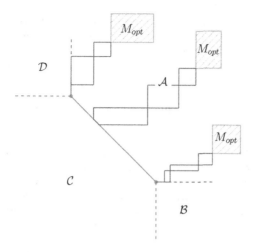

Fig. 9. Intersection of *Pareto cells* with edges of $\mathcal{FVD}$

4   Conclusion and Future Work

In this paper we introduced an important and useful multi-objective optimization problem with 1-median and 1-center in Manhattan metric as its objectives. We considered the problem in one and two dimensional space. We also determined the Pareto optimal front and Pareto set simultaneously. Furthermore we proved finding Pareto front and Pareto solution set of proposed problem is $\theta(n \log n)$.

In higher dimensions, considering Manhattan metric, similar to two dimensional space we can show that optimal of 1-median, i.e. $M(x)$, will be a d dimensional hypercube. Also, it can be computed in $\mathcal{O}(dn)$. For optimal of 1-center, i.e. $C(x)$, the propositions are not straight forward. However finding the smallest circumferential hypercube drives us to the hyperplane which is the locus of cube's center (optimal of 1-center). Moreover, it seems that farthest point Voronoi diagram has the most $2d$ regions. Thus we guess the Pareto optimal set is very similar to two dimensional space; i.e. smallest interval of hypercubes from $M(x)$ to $C(x)$ which are connected by their corners in direction perpendicular to locus of optimal of $C(x)$.

In Euclidean metric, we think this problem will be much harder and the Pareto solutions cannot be computed exactly. In this case we have to approximate Pareto solutions and Pareto front. Moreover, this approximation can be followed for harder objectives such as 2-median and 2-center.

References

1. Arya, V., Garg, N., Khandekar, R., Meyerson, A., Munagala, K., Pandit, V.: Local search heuristics for k-median and facility location problems. SIAM J. Comput. **33**(3), 544–562 (2004)

2. Badoiu, M., Clarkson, K.L.: Smaller core-sets for balls. In: Proceedings of the Fourteenth Annual ACM-SIAM Symposium on Discrete Algorithms, Society for Industrial and Applied Mathematics, pp. 801—802 (2003)
3. Bādoiu, M., Har-Peled, S., Indyk, P.: Approximate clustering via core-sets. In: Proceedings of the Thiry-Fourth Annual ACM Symposium on Theory of Computing, pp. 250–257. ACM (2002)
4. Bartal, Y.: Probabilistic approximation of metric spaces and its algorithmic applications. In: Proceedings of 37th Annual Symposium on Foundations of Computer Science, pp. 184–193. IEEE (1996)
5. Charikar, M., Guha, S., Tardos, É., Shmoys, D.B.: A constant-factor approximation algorithm for the k-median problem. In: Proceedings of the Thirty-First Annual ACM Symposium on Theory of Computing, pp. 1–10. ACM (1999)
6. Chazelle, B., Matoušek, J.: On linear-time deterministic algorithms for optimization problems in fixed dimension. J. Algorithms **21**(3), 579–597 (1996)
7. Coello, C.A.C., Van Veldhuizen, D.A., Lamont, G.B.: Evolutionary Algorithms for Solving Multi-objective Problems, vol. 242. Springer, Verlag (2002)
8. Deb, K.: Multi-objective Optimization Using Evolutionary Algorithms, vol. 16. Wiley, Chichester (2001)
9. Har-Peled, S., Kushal, A.: Smaller coresets for k-median and k-means clustering. In: Proceedings of the Twenty-First Annual Symposium on Computational Geometry, pp. 126–134. ACM (2005)
10. Har-Peled, S., Mazumdar, S.: Coresets for k-means and k-median clustering and their applications, pp. 291–300 (2004)
11. Hochbaum, D., Shmoys, D.: A best possible approximation algorithm for the k-center problem. Math. Oper. **10**, 180–184 (1985)
12. Jain, K., Mahdian, M., Saberi, A.: A new greedy approach for facility location problems. In: Proceedings of the Thiry-Fourth Annual ACM Symposium on Theory of Computing, pp. 731–740. ACM (2002)
13. Megiddo, N., Supowit, K.J.: On the complexity of some common geometric location problems. SIAM J. Comput. **13**(1), 182–196 (1984)
14. Meyerson, A., O'Callaghan, L., Plotkin, S.: A k-median algorithm with running time independent of data size. Mach. Learn. **56**(1–3), 61–87 (2004)
15. Papadimitriou, C.H., Yannakakis, M.: On the approximability of trade-offs and optimal access of web sources. In: Proceedings of 41st Annual Symposium on Foundations of Computer Science, pp. 86–92. IEEE (2000)
16. Shmoys, D.B.: Approximation algorithms for facility location problems. In: Jansen, K., Khuller, S. (eds.) APPROX 2000. LNCS, vol. 1913, pp. 27–32. Springer, Heidelberg (2000)
17. Tansel, B.C., Francis, R.L., Lowe, T.J.: State of the artlocation on net- works: a survey. part i: the p-center and p-median problems. Manage. Sci. **29**(4), 482–497 (1983)
18. Zarrabi-Zadeh, H., Chan, T.M.: A simple streaming algorithm for minimum enclosing balls. In: CCCG. Citeseer (2006)
19. Zarrabi-Zadeh, H., Mukhopadhyay, A.: Streaming 1-center with outliers in high dimensions. In: CCCG, pp. 83–86 (2009)

Author Index

Printed in the United States
By Bookmasters